Thinking Critically: Renewable Energy

John Allen

ReferencePoint
Press®

San Diego, CA

© 2014 ReferencePoint Press, Inc.
Printed in the United States

For more information, contact:
ReferencePoint Press, Inc.
PO Box 27779
San Diego, CA 92198
www. ReferencePointPress.com

Picture Credits:
Cover: Thinkstock Images
Maury Aaseng: 8, 16, 21, 29, 34, 41, 49, 56, 62

LIBRARY OF CONGRESS CATALOGING-IN-PUBLICATION DATA

Allen, John.
  Thinking critically : renewable energy / by John Allen.
    pages cm. -- (Thinking critically)
  Includes bibliographical references and index.
  ISBN-13: 978-1-60152-628-1 (hardback)
  ISBN-10: 1-60152-628-8 (hardback)
 1. Renewable energy resources. I. Title.
  TJ808.A459 2013
  333.79'4--dc23

                                         2013033656

Contents

Foreword   4
Overview: Renewable Energy   6

Chapter One: Are Renewable Energy Sources Needed?
The Debate at a Glance   12
Renewable Energy Sources Are Needed   13
Renewable Energy Sources Are Not Needed   19

Chapter Two: How Practical Is Renewable Energy?
The Debate at a Glance   25
Renewable Energy Is Very Practical   26
Renewable Energy Is Not Practical   32

Chapter Three: Is Renewable Energy Too Expensive?
The Debate at a Glance   38
Renewable Energy Is Too Expensive   39
Renewable Energy Is Affordable   45

**Chapter Four: Should the Government Help
Develop Renewable Energy?**
The Debate at a Glance   51
The Government Should Help Develop Renewable Energy   52
The Government Should Not Help Develop Renewable Energy   58

Source Notes   64
Renewable Energy Facts   68
Related Organizations and Websites   71
For Further Research   74
Index   76
About the Author   80

Foreword

"Literacy is the most basic currency of the knowledge economy we're living in today." Barack Obama (at the time a senator from Illinois) spoke these words during a 2005 speech before the American Library Association. One question raised by this statement is: What does it mean to be a literate person in the twenty-first century?

E.D. Hirsch Jr., author of *Cultural Literacy: What Every American Needs to Know*, answers the question this way: "To be culturally literate is to possess the basic information needed to thrive in the modern world. The breadth of the information is great, extending over the major domains of human activity from sports to science."

But literacy in the twenty-first century goes beyond the accumulation of knowledge gained through study and experience and expanded over time. Now more than ever literacy requires the ability to sift through and evaluate vast amounts of information and, as the authors of the Common Core State Standards state, to "demonstrate the cogent reasoning and use of evidence that is essential to both private deliberation and responsible citizenship in a democratic republic."

The Thinking Critically series challenges students to become discerning readers, to think independently, and to engage and develop their skills as critical thinkers. Through a narrative-driven, pro/con format, the series introduces students to the complex issues that dominate public discourse—topics such as gun control and violence, social networking, and medical marijuana. All chapters revolve around a single, pointed question such as Can Stronger Gun Control Measures Prevent Mass Shootings?, or Does Social Networking Benefit Society?, or Should Medical Marijuana Be Legalized? This inquiry-based approach introduces student researchers to core issues and concerns on a given topic. Each chapter includes one part that argues the affirmative and one part that argues the negative—all written by a single author. With the single-author format the predominant arguments for and against an issue can be synthesized into clear, accessible discussions supported by details and evidence including relevant facts, direct quotes, current examples, and statistical illustrations. All volumes include focus questions to guide students as they read each pro/con discussion, a list of key facts, and an annotated list of related organizations and websites for conducting further research.

The authors of the Common Core State Standards have set out the particular qualities that a literate person in the twenty-first century must have. These include the ability to think independently, establish a base of knowledge across a wide range of subjects, engage in open-minded but discerning reading and listening, know how to use and evaluate evidence, and appreciate and understand diverse perspectives. The new Thinking Critically series supports these goals by providing a solid introduction to the study of pro/con issues.

Renewable Energy

In March 2010 US president Barack Obama toured the manufacturing plant of a Silicon Valley solar-energy firm called Solyndra. In remarks following his tour, Obama praised the company's plans to hire one thousand new workers and produce enough solar panels to replace the power generated by eight coal-fired electricity plants.

Eighteen months later, Solyndra's prospects had collapsed in failure. Its innovative but expensive solar panels could not compete with a flood of cheaper panels from China. The company declared bankruptcy, closed its factory, and laid off most of its workers. News reports focused on the more than $500 million in stimulus funds Solyndra had received from the federal government. Debates raged about the wisdom of the federal government trying to pick winners among companies in the fast-moving energy and technology sectors. *Solyndra* became shorthand for the pitfalls faced by renewable energy companies.

Nevertheless, environmentalists and green supporters were undaunted. Some questioned Solyndra's business model, while others pointed out that the media focused on alternative energy flops and virtually ignored the success stories—such as NRG Energy's California Valley Solar Ranch project, a huge solar installation in central California that is set to generate emission-free power for more than one hundred thousand homes. Supporters insisted that setbacks such as Solyndra's did not change the fact that renewable energy sources were urgently needed to replace fossil fuels. "Contrary to all the chatter out there, solar's here, solar's ready, solar's booming,"[1] said Arno Harris, the CEO of a competing solar firm. Lisa Murkowski, a Republican senator from Alaska, warned against hasty reactions to Solyndra's failure and expressed her support for federally backed renewable energy programs.

What Is Renewable Energy?

Renewable energy comes from resources that are renewed in nature, such as water, wind, sunshine, and geothermal. These permanent sources of energy cannot be depleted by overuse. Renewable energy is often called green energy or clean power because it tends to result in much less pollution of the air or water. Renewable energy is also referred to as alternative energy—meaning alternative to fossil fuels.

Fossil fuels include coal, oil, and natural gas. These materials formed deep below the earth's surface over millions of years due to the decomposition of animal and plant matter. Industrial economies all over the world depend mostly on fossil fuels for their energy needs. Yet the burning of fossil fuels in cars, homes, and factories not only causes air pollution from smoke and exhaust fumes but also contributes to higher levels of carbon dioxide in the atmosphere. In addition, environmentalists point out the damage to natural ecosystems that can result from harvesting fossil fuels, including disasters like the 2010 BP oil spill in the Gulf of Mexico.

Why Is Renewable Energy Important?

Aside from helping reverse the damaging effects of fossil fuels, renewable energy is also needed to satisfy the world's growing appetite for energy. Industrialized nations consume energy at an astounding pace, and as poorer nations develop, they also require more energy, adding to the burdens on the planet's supplies. According to the US Department of Energy, total world energy demand each year is about 400 quadrillion British thermal units (BTUs). (One BTU is roughly equivalent to the heat and energy of a struck match.) By the year 2020 this number is expected to increase by 50 percent. Currently, fossil fuels such as coal, oil, and natural gas make up about 87 percent of total energy use, supplying more than 350 quadrillion BTUs a year. As fossil fuel reserves are gradually depleted, renewable technologies such as wind, solar, hydrothermal, and geothermal power represent a great opportunity: to supply the world with sources of clean, reliable energy that can never be used up.

An even more important reason to adopt renewable energy, environmentalists say, is the need to prevent an oncoming climate disaster.

Percentage of Electricity from Renewable Sources

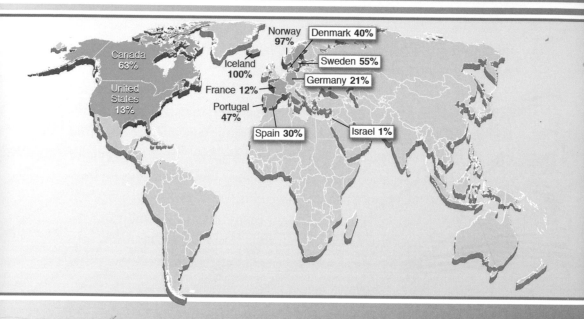

Source: *The New York Times*, "How Much Electricity Comes from Renewable Sources," March 23, 2013. www.nytimes.com.

Most scientists today blame rising levels of carbon dioxide in the atmosphere for climate change or global warming. This warming occurs by a natural process called the greenhouse effect. Radiant heat from the sun is trapped in the earth's lower atmosphere, keeping the surface of the earth relatively warm. Because industrialized nations are using more coal, oil, and natural gas for energy, levels of atmospheric carbon dioxide continue to rise. Scientists have theorized that this increase of heat-trapping gas in the atmosphere is the main contributor to global warming. In fact, NASA climate scientist James Hansen said that the tremendous amount of new heat energy trapped in the atmosphere is "equivalent to exploding 400,000 Hiroshima atomic bombs per day 365 days per year."[2]

Concerns about global warming have led governments to adopt policies to reduce greenhouse gas emissions. In 1997 many industrialized nations (not including the United States) signed a United Nations–

sponsored treaty called the Kyoto Protocol that required the signatories to reduce their collective greenhouse gas emissions by 5.2 percent. Separately, nations have pursued various methods to reduce carbon emissions, from shutting down coal-fired plants and setting new standards for automobile emissions to offering subsidies for renewable energy businesses.

Critics of these governmental policies also tend to be skeptical of the whole idea of global warming. Recently these "climate skeptics" have pointed to reports indicating that little or no warming has occurred in the last ten to fifteen years, despite a massive increase of greenhouse gases in the atmosphere. Still, environmentalists counter that this short-term plateau does not negate the overall warming trend or the evidence of polar ice erosion and rising sea levels, and they assert that most of the warmest years on record have occurred recently. Thus, the debate about climate change is closely connected to the debate about the need for renewable energy.

Technologies That Produce Renewable Energy

Whether as a safeguard against energy shortages or a solution to climate change, renewable energy must prove its worth as a practical source of power. Research continues to improve the technologies used to collect, store, and distribute renewable energy. The basic technologies include the following.

Solar power: Photovoltaic (PV) technology converts energy from sunlight into electricity. A PV cell is made up of two or more thin layers of semiconducting material, usually silicon. When exposed to sunlight, the silicon generates electrical charges that can be conducted away by metal contacts as direct current. Since the electrical output from each PV cell is small, multiple cells are linked together and enclosed (commonly behind glass) to make a module, or solar "panel." Any number of PV modules can be connected into a system to produce the desired output of electricity. When linked to a local electricity network, a PV system's solar electricity can either be used at once (as with homes or commercial buildings) or sold to an electric utility, whose grid serves as a storage system.

Wind power: Wind turbines harness the power of wind to generate electricity. Most wind turbines have three blades. As air rushes past the blades, it creates lift, causing the blades to rotate. A drive shaft connected to the blades turns the motor of a generator, which produces electricity. Grouped together in wind farms on plains, hills, and offshore locations, these wind turbines (or windmills) are able to provide electric power to entire neighborhoods or contribute electricity to a local power grid. One huge turbine in a windy area can produce enough electricity to power more than fourteen hundred homes.

Water power: Hydroelectric power is currently the most widely used form of renewable energy, accounting for nearly 20 percent of the world's electricity and more than 80 percent of the energy produced from renewable sources. Hydroelectric power harnesses the force of moving water to generate electricity. In a hydroelectric facility, water is stored behind a dam to form a reservoir, or artificial lake. The force of the water as it is released from the reservoir through the dam turns the blades of a giant turbine. As with wind power, the drive shaft of the turbine blades activates the motor of a generator to produce electricity.

Geothermal power: In geothermal technology, steam from underground hot springs is funneled directly into a turbine. The energy of the steam spins the turbine blades, which turn a shaft connected to a generator. Many geothermal fields use 95 percent of the available energy, the highest percentage among renewable sources.

Biomass energy: Biomass technology is the burning of waste material to make electricity. The matter that makes up biomass includes everything from tree branches, grass clippings, wastepaper, and crop residues to used tires, sawdust, and manure. These waste products are gathered in large trucks and hauled to a biomass power plant. There the biomass is emptied into large hoppers and fed into a large furnace. Heat produced by the burning biomass boils water to create steam, which turns the blades of a turbine connected to a generator.

These are only some of the promising technologies for producing renewable energy. Scientists and technicians are constantly working to make renewable sources more practical and cost-competitive. Perhaps some renewable energy source yet unheard of will prove to be the ultimate solution to the world's energy needs. As the quest for reliable and abundant power generation continues, it seems certain that renewable energy in many different forms will remain an important part of the debate.

Chapter One

Are Renewable Energy Sources Needed?

Renewable Energy Sources Are Needed

- Unlike fossil fuels, renewable energy sources are unlimited in supply and will never be depleted.
- Renewable energy sources are cleaner and therefore better for the environment than are fossil fuels.
- As nations grow, they need to make use of every possible energy source to meet their increasing need for power.
- Focusing on renewable energy sources will enhance energy security worldwide.

The Debate at a Glance

Renewable Energy Sources Are Not Needed

- New technologies are serving to increase the amount of recoverable fossil fuels available for use.
- Scientists are working on synthetic fuels that are more promising than renewable energy sources.
- Exploiting renewable energy sources requires a large amount of fossil fuel energy.
- Renewable energy sources have their own environmental disadvantages.

Renewable Energy Sources Are Needed

"If we don't act, if we don't do anything, if we don't invest anything, we can be sure that we will have a catastrophe very soon. We have to have confidence to invest in the new energy. We can act together to create this world of renewable energy."

—French president François Hollande addressing the World Future Energy Summit on January 15, 2013.

Quoted in Adam Vaughan, "François Hollande: Invest in Renewable Energy to Avert 'Catastrophe,'" *Guardian* (London), January 15, 2013. www.theguardian.com.

Consider these questions as you read:

1. When you consider the facts and assertions in this discussion, how persuasive is the case that renewable energy sources are needed? Which facts and ideas did you find most persuasive?
2. Which renewable energy source is most promising as a long-term energy solution? Explain your answer.
3. Should the environmental impact of an energy source be the most important consideration in the debate over its use? Why or why not?

Editor's note: The discussion that follows presents common arguments made in support of this perspective, reinforced by facts, quotes, and examples taken from various sources.

While fossil fuels continue to supply most of the world's energy needs—the world consumes about 150 million gallons (568 million L) of oil every hour—people are increasingly looking for other sources of energy for the future. Many see an urgent need to focus on renewable sources of energy: sun, wind, geothermal, hydropower, and biomass. According to Steven Koonin, director of the Center for Urban Science and Progress at New York University, "There are compelling reasons to improve our energy system—to make it more accessible, affordable and reliable, and to reduce its environmental impact."[3] A Gallup poll in March 2013 showed that, despite the continuing use of fossil fuels, Americans support more development of renewable energy. In the poll, about three-fourths of Americans favored more emphasis on solar power and 71 percent on wind power. By contrast, only 46 percent thought more emphasis should be put on oil, 37 percent on nuclear power, and 31 percent on coal. "This is how the typical American thinks in 2013," said environmental writer John Upton. "'More solar power, please. No more nuclear, thanks though. And let's get ready for this crazy climate-change thing.'"[4]

Citizens are putting this belief into practice around their own homes by installing rooftop solar panels. The Solar Energy Industries Association says about fifty-two thousand residential rooftop systems were installed in 2011, and total commercial and private installations were projected to rise by about 70 percent in 2012.

Unlimited in Supply

One reason for changing opinions about energy use might be the realization that fossil fuels are finite. No matter how much oil, natural gas, or coal is discovered, these resources will run out someday. Every new oil field or natural gas well merely delays the inevitable. By contrast, the power contained in sunlight, rushing wind, or flowing water can never be used up. According to *American Energy Independence*, an Internet magazine that advocates for renewable energy, "Every hour, the sun radiates more energy onto the earth than the entire human population uses in one whole year." The magazine goes on to point out that "on average, and particularly in the Sunbelt regions of the Southwestern United States, every square meter area exposed to direct sunlight will receive about 1 kilowatt-hour per hour of solar energy."[5] Even allowing that the most productive sun angles occur only for about six hours a day, this represents a huge amount of potential energy for generating electricity and heating houses and water. In addition, engineers have theorized that satellites in space could someday collect solar energy and transmit it wirelessly to earth in an uninterrupted flow.

Wind and geothermal energy, hydropower, and biofuels all have equally enormous potential for supplying future energy needs. Companies like GE Wind Energy in the United States, the world's largest supply and service company for wind turbines, are committed to tapping this vast power source by building wind farms that extend for miles. A study by the US Department of Energy concluded that wind power could supply 20 percent of US electricity by 2030. As for hydropower, it is currently the largest source of renewable energy in the United States, and its potential is much greater still. Only about twenty-five hundred of the country's eighty-three thousand dams and water-diversion facilities are actually used to generate power. The Energy Department estimates that there are more than twenty-five thousand megawatts of undeveloped capacity (one megawatt is sufficient to power about 650 houses). Hydropower's potential is being realized in less-developed areas as well. It is used to generate more than 85 percent of electricity in Ethiopia, for example. Ethiopia and Kenya are also seeing large investments in geothermal energy, which is a limitless source of underground heat. Finally, trees and crops used to make biomass and biofuels can be regrown as long as there is sufficient soil and water. Facts like these indicate the unlimited potential of renewables.

> "Every hour, the sun radiates more energy onto the earth than the entire human population uses in one whole year."[5]
>
> —*American Energy Independence*, an Internet magazine that advocates for renewable energy.

Cleaner and Better for the Environment

One of the best aspects of renewable energy is that it is also cleaner energy. In general, while all energy sources impact the environment, renewable sources do much less harm. This includes most measures of environmental impact: air and water pollution, public health effects, wildlife and habitat damage or loss, effects on water and land use, and emissions related to climate change. Use of fossil fuels has increased the level of carbon dioxide in the atmosphere by more than 25 percent over the last century and a half. Climate scientists forecast that a continued increase in these levels of carbon dioxide and other gases will result in significant

Americans Favor Increased Use of Alternative Energy

A March 2013 Gallup poll asked Americans which sources of energy should be emphasized more in the United States. The results showed that a sizable majority of Americans favored more emphasis on renewable sources to produce domestic energy. Republican support was less than that of Democrats and Independents but still registered well over 50 percent.

Do you think that as a country, the United States should put more emphasis, less emphasis, or about the same emphasis as it does now on producing domestic energy from each of the following sources?

US Should Place "More Emphasis" on Each Source of Domestic Energy Production, by Party ID

	All Americans	Republicans	Independents	Democrats
Solar power	76%	68%	74%	87%
Wind	71%	59%	68%	83%
Natural gas	65%	78%	62%	59%
Oil	46%	71%	43%	29%
Nuclear power	37%	49%	35%	30%
Coal	31%	51%	26%	21%

Source: Gallup, "Americans Want More Emphasis on Solar, Wind, Natural Gas," March 27, 2013. www.gallup.com.

warming of the atmosphere, leading to melting glaciers and rising sea levels, altered weather patterns and more extreme weather events, more frequent droughts, and other climate-related problems. Renewable energy sources such as solar power, wind power, and geothermal energy do not add gases to the atmosphere and thus represent a solution to global warming concerns.

Many countries are beginning to realize that using renewable energy also leads to cleaner air and water. China, which now burns 3.8 billion tons (3.4 billion metric tons) of coal each year, an amount nearly equal to the rest of the world combined, is facing an environmental crisis related to its fossil fuel use. Journalists have dubbed the shrouds of fog that cover the capital city of Beijing "the airpocalypse." The blanket of pollution is so thick it is visible from space. Bloggers in the city regularly post their complaints about the wretched air quality. "I love our city, but I refuse to be a human vacuum cleaner,"[6] wrote one disgruntled Beijing blogger. The polluted air has been shown to increase respiratory infections, heart disease, and lung cancer. On January 23, 2013, Chinese political and business leaders formed the Clean Air Alliance of China. They vowed to reduce emissions and focus more on using clean, renewable energy sources such as solar and wind power.

> **"I love our city, but I refuse to be a human vacuum cleaner."[6]**
>
> —A Chinese blogger in Beijing responding to air quality in the city.

Meeting the Growing Need for Energy

Concerns about the environmental costs of fossil fuels coincide with a tremendous increase in demand for energy worldwide. It seems inevitable that the clash of these two factors will work in favor of renewable energy. According to forecasts by the US Energy Information Administration (EIA), world energy consumption will increase by 53 percent between 2008 and 2035. Most of this explosive growth will occur in emerging economies, where energy demands are projected to grow by 117 percent. "Rising prosperity in China and India is a major factor in the outlook for global energy demand," said Adam Sieminski, head of the EIA. Sieminski pointed out that the two countries combined will account for half the world's total increase in energy use through 2040. "The question is how do we accommodate rising prosperity and still maintain energy security and the environment?"[7] Sieminski said. Trying to meet this increasing demand entirely with fossil fuel production can only worsen problems of pollution and carbon emission. Seeing this, governments are already instituting supports for alternative, cleaner energy sources. For example, the Barack Obama administration included an extensive package of renewable energy subsidies and tax breaks in its 2009 economic stimulus bill.

Enhanced Energy Security

Rising energy demand helps emphasize a fact of life in geopolitics: The world's fossil fuel resources are not distributed evenly. Some nations have vast reserves of oil and natural gas while: others must import almost all the fuel they need. This disparity in energy resources can create a high level of tension between the energy haves and have-nots, particularly when the haves manipulate the supply in response to political events. In 1973 the Organization of the Petroleum Exporting Countries (more often called OPEC) oil cartel reacted to US support for Israel in the Yom Kippur War by raising oil prices steeply through means of an embargo, resulting in an oil crisis that saw fuel prices explode in many Western nations. Thirty years later the US-led invasion of Iraq interrupted the sale of Iraqi oil and led to much higher oil prices worldwide. President George W. Bush responded to this problem in 2007 by signing the Energy Independence and Security Act to promote production of clean, renewable fuels and reduce dependence on foreign sources of energy.

In today's interconnected world, accidents or energy shortages in one country can produce ripple effects that are impossible to predict. Many politicians now routinely speak of energy security, linking the ideas of national security and access to the sort of reliable sources of power that renewable energy can provide. With renewable energy, which is clean, abundant, and widely available, the nations of the world have an opportunity to fundamentally change the way they generate and use power. That is why renewable energy is so urgently needed today and will be even more so in the future.

Renewable Energy Sources Are Not Needed

"We rushed into renewable energy without any thought. The schemes are largely hopelessly inefficient and unpleasant. I personally can't stand windmills at any price."

—James Lovelock, a British scientist and environmental author.

Quoted in Leo Hickman, "James Lovelock: The UK Should Be Going Mad for Fracking," *Guardian* (London), June 15, 2012. www.theguardian.com.

Consider these questions as you read:

1. Do you agree that recent improvements in ways to extract fossil fuels reduce the need for renewable energy? Why or why not?
2. What developments must occur in synthetic fuels to make them viable as an alternative to renewables and fossil fuels? Do you think you will see these developments in your lifetime? Explain.
3. How would you respond if a solar or wind farm were being installed near your neighborhood? What objections might you have? Should citizens have a voice in how power generation is achieved? Explain.

Editor's note: The discussion that follows presents common arguments made in support of this perspective, reinforced by facts, quotes, and examples taken from various sources.

Far from being nearly played out, as the green movement claims, fossil fuel supplies continue to expand. This is mostly due to new technologies that allow for the capture of fossil fuels that were either not accessible or not economically feasible to go after just a few years ago. As recently as 2012 the EIA reported huge increases in proven reserves of crude oil and natural gas in the United States. For example, natural gas reserves grew by 11 percent in one year—and this was the seventh straight year of increases—while crude reserves rose by 9 percent. Proven reserves are energy supplies that are charted in detail and could be tapped under today's market conditions.

Total recoverable reserves in the United States are doubtless far greater, and the last few years have only seen more increases in predicted reserves. In the EIA report, its authors declared, "These increases demonstrate the possibility of an expanding role for domestic natural gas and crude oil in meeting both current and projected U.S. energy demands"[8]—not exactly the picture of fossil fuel decline painted by many environmentalists. In addition, an updated EIA report on world shale gas resources in 2013 declared that these reserves had risen by 47 percent.

The truth is that the United States and many other countries are experiencing explosive growth in recoverable supplies of crude oil and natural gas, and special technologies are driving much of this growth.

> "These increases demonstrate the possibility of an expanding role for domestic natural gas and crude oil in meeting both current and projected U.S. energy demands."[8]
>
> —Report from the EIA.

Hydraulic fracturing, or fracking, is a method of opening fractures in rock formations by injecting cracks with a mixture of sand and water. The larger fissures that result enable more oil and gas to flow out of the rock formation and into the well bore. Fracking has made many promising shale formations economically viable for the first time because of the increased levels of extraction. According to Dieter Helm, an energy expert at Oxford University, "In the U.S., shale gas didn't exist in 2004. Now it represents 30% of the market."[9] Another new procedure is the farming of oil sands. Found primarily in northern Alberta, Canada, and in certain parts of Venezuela, oil sands are areas of sand and rock that contain crude bitumen, a heavy, viscous type of crude oil. The bitumen in oil sands is extracted by mining or steam injection and then processed into gasoline and other fuels.

The US Energy Information Administration's annual survey of
domestic reserves of oil and natural gas has shown significant
increases in recent years, suggesting little need for a switch to
renewable energy sources. Such a switch would likely be both
time consuming and costly.

US Oil and Natural Gas Proved Reserves, 1981–2011

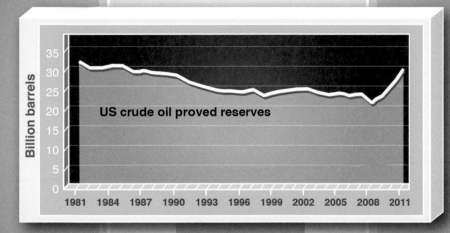

US crude oil proved reserves

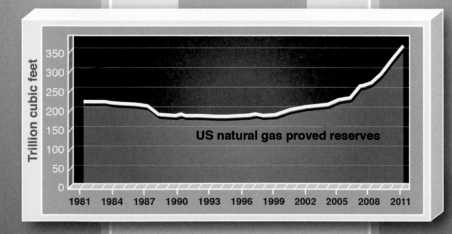

US natural gas proved reserves

Source: US Energy Information Administration, "US Crude Oil and Natural Gas Proved Reserves," August 1, 2013.
www.eia.gov.

Promising Synthetic Fuels

Owing to the abundance of oil for making gasoline and diesel, the world has come to depend on liquid fuels for internal combustion engines. Aside from ethanol and biodiesel, which are inefficient fuel additives, renewable energy sources are used to produce electricity, not liquid fuel. Electric-powered motors for cars and trucks are much less practical than gasoline-powered engines due to problems with battery storage and charging times. Also, in order to accommodate electric vehicles, tens of thousands of charging stations would have to be installed across the nation and throughout the world. By contrast, clean-burning synthetic fuels could one day replace fossil fuels almost seamlessly, enabling motorists to fill their tanks at service stations in the customary way.

There are promising signs of future success for synthetic fuels. For example, engineers at a small British company have produced synthetic gasoline by extracting carbon dioxide from air and hydrogen from water and then combining them in a reactor. "It's actually cleaner because it's synthetic," said Peter Harrison, the CEO of Air Fuel Synthesis. "You just make what you need to make in terms of the contents of it, so it doesn't contain what might be seen as pollutants, like sulphur."[10] The company's engineers are confident that the process can be scaled up to meet refinery-size needs in the future.

Another synthetic fuel is made from genetically engineered bacteria. John Love leads a team of researchers at the University of Exeter that is developing a biofuel that will work in modern engines. "For an engine to function optimally, it needs high-quality fuel—fossil fuels offer that very high quality," Love said. "The challenge we faced was finding a way to make the fuel that the retail industry needs biologically, rather than mining it from the ground."[11] If successful, these synthetic fuel projects will move beyond today's impractical additives and electric motors to provide an unlimited supply of energy.

Renewables Require Fossil Fuels

Just as the case for renewable fuel additives can be misleading, so is the idea that renewable energy requires little or no input from fossil fuels. On the contrary, tapping into renewable sources such as sun or wind power requires quite a lot of fossil fuel energy. For example, take solar panels. A little-known truth about these environmental wonders is that the power used to make them comes mainly from fossil fuel–generated electricity. In fact, a study by Stanford University researchers suggests that only by 2013 had the total generating capacity of solar panels around the world made up for the energy that went into manufacturing all those panels since 2000. The study sees this as an important turning point for the solar industry. However, the researchers' optimism is based on the notion that installation and material costs will continue their rapid drop in the next decade, which is no sure thing.

> "Our results suggest that existing biofuel policies have been very costly, produce negligible reductions in fossil fuel use and increase, rather than decrease, greenhouse gas emissions."[12]
>
> —Bill Jaeger, lead author of an Oregon State University study on the cost-effectiveness of biofuels.

Similarly, wind power relies heavily on fossil fuels. In countries like Germany, where wind power is a major component of the power grid, the energy needed to build and install wind turbines and connect them to the power network is significant and comes mainly from fossil fuels. This is in addition to the energy required as a backup when wind supply at power stations is insufficient.

Finally, a 2011 study by economists at Oregon State University questions the rationale for biofuels such as ethanol. The study emphasizes that biofuels require fossil fuels for their production and transportation. For example, the corn used to make ethanol is grown with nitrogen fertilizer, which is made using natural gas. "Our results suggest that existing biofuel policies have been very costly, produce negligible reductions in fossil fuel use and increase, rather than decrease, greenhouse gas emissions,"[12] said Bill Jaeger, lead author of the study. Time and again, a closer look at the processes of producing renewable energy—whether wind, sun, or biomass—leads to the conclusion that they scarcely cut fossil fuel use and are an enormous waste of resources.

Environmental Drawbacks of Renewables

A closer look at environmental impacts also yields interesting results. Green advocates are quick to point out the environmental problems with fossil fuels: Burning them pollutes the air and water; extracting them can result in damage to land, water, and animal habitats; and their emissions contribute to global warming. Yet supposedly pristine renewable fuels have their own environmental baggage. As renewable sources become a larger part of the energy picture, people are learning firsthand how they too affect the environment.

In China, where production of solar panels is often poorly regulated, chemical by-products from their manufacture routinely are spilled into rivers and lakes or spewed into the air. A trace gas such as nitrogen trifluoride, which is seventeen thousand times more concentrated than carbon dioxide, is released during the manufacture of certain solar panels. Most panels have an estimated life span of about twenty-five to thirty years, after which they become waste that is potentially hazardous due to their containing toxic substances such as cadmium. Environmentalists expect the solar industry to take responsibility for used panels, but that remains uncertain. "Some companies have already signed up for it," said Vasilis Fthenakis, head of the Photovoltaic Environmental Research Center at Brookhaven National Laboratory. "Some others were skeptical, some others don't think it's necessary yet. We hope the industry will adopt recycling."[13]

Wind, hydropower, and geothermal power also have environmental drawbacks. Most wind turbines are very large and noisy to operate. Many people consider a long line of wind turbines on the horizon to be an eyesore, and the turning blades are a constant threat to bird populations. Environmental groups also object to hydroelectric dam projects because dams alter the natural flow of rivers and threaten marine habitats. Even geothermal energy has its ecological downside, because drilling reservoirs can bring up hazardous chemicals along with the hot water. The moral to the renewable energy story is that there is no "free lunch," environmentally speaking, in the quest for new sources of power.

Chapter Two

How Practical Is Renewable Energy?

Renewable Energy Is Very Practical

- There is no need to "find" renewable energy sources, since they exist everywhere.
- Certain renewable energy technologies solve the problem of intermittent power.
- Wind power is especially practical in areas like the southern plains, where winds are almost constant.
- Researchers are working on ways to integrate national power grids for more renewable energy use.

The Debate at a Glance

Renewable Energy Is Not Practical

- Constraints due to land capacity limit the share of a population's energy needs that renewable sources can provide.
- Sunlight and wind are too intermittent to be practical power sources in many areas.
- Technology for storing energy as electricity is still lacking, so alternative sources cannot maintain the power grid.

Renewable Energy Is Very Practical

"We have the tools to significantly ramp up renewable energy use and keep the lights on. With ingenuity, innovation, and smart policies, we can fully transition to a clean, renewable electricity system."

—Union of Concerned Scientists, an organization devoted to environmental solutions.

Quoted in Union of Concerned Scientists, "Ramping Up Renewables," April 2013. www.ucusa.org.

Consider these questions as you read:

1. Which form of renewable energy shows the most promise for becoming a major practical energy source? Why?
2. Why is development of the "smart grid" so important to the future of renewable energy?
3. Taking into account the facts and ideas in this discussion, are you persuaded that renewable energy is a practical alternative to fossil fuels? Which facts and ideas are most persuasive, and why?

Editor's note: The discussion that follows presents common arguments made in support of this perspective, reinforced by facts, quotes, and examples taken from various sources.

Renewable energy is gaining acceptance among members of the public because it is increasingly seen as a practical alternative to fossil fuels. Recent polls consistently show that Americans prefer a focus on renewable sources such as solar and wind power to oil and coal. While in the past an economy based on renewable energy seemed a mere fantasy, today it is a topic for serious debate. According to environmental writer Chris Williams, "It may seem hard to believe, but it is fully within our means today to make the alternative energy dream a green reality. All the technologies exist. The engineering is relatively straightforward, especially when compared to the epic size of our oil-powered, automobile-based societies. The need is obvious."[14] He goes on to assert that politics more than technology is preventing greater use of renewables.

Some researchers are trying to demonstrate just how practical renewable energy is. Mark Jacobson and a team of scientists at Stanford University have developed a detailed plan to power the state of New York entirely with wind, water, and solar energy by 2030. For example, the plan specifies the number of wind turbines, solar-receptive rooftops, and geothermal plants required. "If society is going to do it," Jacobson declares, "at least we now know that it's technically and economically feasible. Whether it actually happens depends on political will."[15]

Natural Abundance

Practical plans like Jacobson's are made possible by the natural abundance of renewable energy. When an oil company discovers a new field of oil or natural gas, the find is often front-page news because of the assumed scarcity of fossil fuels. With renewable energy, such costly and time-consuming exploration is not necessary. Renewable energy is found everywhere, just waiting to be harnessed. As long as the sun exists, its energy will blanket the earth. Uneven heating of the earth's surface will always produce wind currents. The heat at the earth's core perpetually release geothermal energy.

> "If society is going to [adopt renewable energy], at least we now know that it's technically and economically feasible. Whether it actually happens depends on political will."[15]
>
> —Mark Jacobson, a researcher at Stanford University.

Abundance is one of the great practical advantages of renewable energy. While fossil fuels are being used up every day, the prospects for endlessly renewable energy continue to grow. US Census Bureau data indicate that solar receptors could be installed on about 100 million rooftops in the United States, generating more than 90 percent of the kilowatt-hours (kWhs) of electricity now produced by all the nation's sources. A 2012 study by the Carnegie Institution for Science makes an even stronger case for wind power. The study suggests that all of the world's energy needs could be supplied by harnessing surface and high-altitude winds. As environmental writer Mark Halper notes, "There's enough wind blowing out there to provide more than 100 times the amount of electricity that the world currently uses."[16]

Some experts see similar possibilities for geothermal heat. According to Norway's Are Lund, a senior researcher at SINTEF Materials and Chemistry, "If we can drill and recover just a fraction of the geothermal heat that exists, there will be enough to supply the entire planet with energy—energy that is clean and safe."[17]

Uninterrupted Power Generation

A frequent objection to renewable energy is that its sources are intermittent, or not reliably continuous. The sun does not shine at night, and the wind blows only on certain days. These concerns will diminish with the introduction of new technologies such as concentrated solar power, which provides consistent power even at night or during cloudy weather. Unlike photovoltaic (PV) technology, which converts sunlight directly to electricity via the semiconductor materials in solar panels, a concentrated solar power plant concentrates the sun's energy with mirror devices and then captures it in a heat-transfer fluid. The fluid is used to make steam to drive conventional turbine generators and produce electricity.

Some researchers suggest that the ultimate solution for intermittency is to tap as many sources of renewable energy as possible—mainly onshore and offshore wind and solar—and then employ fossil fuel sources as a backup supply. "In the next few decades new electric power in the U.S. and elsewhere in the world will be dominated by a marriage of wind and solar systems that are intermittent in nature with electrical storage and natural gas–fired systems that provide 'firming' support for the renewable technologies,"[18] said Mark Hannifan, vice president of development at TradeWind Energy, a wind power company.

A study by the University of Delaware found that a sufficiently large system of renewable energy generators could one day meet the demands of a large power grid 99.9 percent of the time. The key is to greatly increase the number of renewable energy generators instead of increasing power storage.

States with More than 60 Percent Energy from Renewable Sources

Successful policies and regional advantages allow certain states to get more than 60 percent of their energy from renewable sources. These sources include wind, solar, water, geothermal, and biomass. As can be seen on the map, several states get 100 percent.

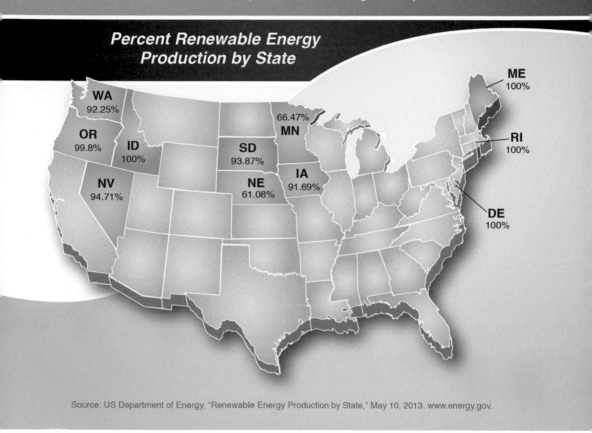

Percent Renewable Energy Production by State

WA 92.25%
OR 99.8%
ID 100%
NV 94.71%
SD 93.87%
NE 61.08%
MN 66.47%
IA 91.69%
ME 100%
RI 100%
DE 100%

Source: US Department of Energy, "Renewable Energy Production by State," May 10, 2013. www.energy.gov.

Willett Kempton, a professor and coauthor of the study, said, "In our 99.9 percent scenario, we found that, in four years, only five times would you need to bring fossil fuel plants back online to ensure power supply."[19]

Regional Advantages

Some regions have obvious practical advantages for supplying renewable energy. California, Nevada, and Arizona, for example, with their abundant sunshine and open spaces, are ideal for large solar panel installations. Solar energy can also provide power in remote areas with little existing infrastructure. The gusty plains of Kansas and Oklahoma are particularly suitable for wind farms. The world's most productive geothermal fields are found in the western states, with the largest located just north of San Francisco, California. These regional sources of renewable energy provide a stable supply of power that can never be used up. In addition, state and local governments find that using renewable energy leads to cleaner air and water and makes areas more attractive for new residents, businesses, and tourists. Fossil fuels, on the other hand, can destroy the tourist trade, as shown by the 2010 *Deepwater Horizon* oil spill that disrupted tourism along the Gulf Coast.

> "There's enough wind blowing out there to provide more than 100 times the amount of electricity that the world currently uses."[16]
>
> —Mark Halper, a writer on environmental issues.

Developing the Smart Grid

To make renewable energy technologies more practical, electricity grids are being updated to twenty-first-century standards. A power grid is the national network that carries electricity from generation plants to consumers, and it consists of wires, substations, switches, and transformers. Creating a "smart grid" means computerizing the system to include sensors that gather data all along the network. A smart grid also allows for centralized control of the thousands or millions of devices in the network. This can help integrate renewable sources of electricity like solar and wind into the grid, whether from large collecting stations or individual homes. Digital sensors can then regulate the flow of power from various sources to ensure a steady supply in the network, with no power surges (too much power) or brownouts (too little). For example, the National Center for Atmospheric Research (NCAR) has worked with Xcel Energy to develop a sophisticated wind-energy forecasting system. The new system allows utility operators to anticipate the amount of energy produced by wind farms in different weather conditions across a service area.

NCAR is also working on a prototype system to create thirty-six-hour solar energy forecasts. The solar-energy forecasting system helps utilities anticipate cloud cover and atmospheric particles that affect the collection rates of solar panels in a service area. "It's critical for utility managers to know how much sunlight will be reaching solar energy plants,"[20] said Sue Ellen Haupt, lead scientist on the NCAR solar project. Detailed forecasts of cloud cover and sunlight intensity are a vital step.

The Barack Obama administration has promoted smart grid technology with public grants worth $3.4 billion. The administration has also overseen the launch of a new Energy Systems Integration Facility devoted to research on how renewable energy can be integrated into the national grid. "Going forward," says John Holdren, the White House director of science and technology, "the Administration will continue to work with states, the electric sector, and other stakeholders to modernize grid infrastructure, develop new tools for a clean energy economy, empower customers, and foster innovation."[21]

Renewable Energy Is Not Practical

"How do you keep the power on when the wind won't blow and the sun won't shine? As California has discovered, it's a problem that's neither easy nor cheap to solve in the brave new world of renewables."

—*Investor's Business Daily*, editorial board.

Quoted in *Investor's Business Daily*, "Green Energy Has a Brownout Problem," editorial, February 27, 2013. http://news.investors.com.

Consider these questions as you read:

1. Do you agree with the contention that renewable energy is not practical because renewable sources cannot be scaled up sufficiently to meet the world's energy needs? Explain.
2. How persuasive is the argument that renewable energy sources require too much land to be practical? Explain.
3. Imagine that scientists discovered a practical and inexpensive method for storing electrical power. How would this affect the debate about fossil fuels versus renewable energy?

Editor's note: The discussion that follows presents common arguments made in support of this perspective, reinforced by facts, quotes, and examples taken from various sources.

Renewable energy sources will never replace fossil fuels for one practical reason: They cannot meet the world's growing energy needs. Renewables are—and probably will remain—a small piece in the global energy puzzle. There is no shortage of pundits who have come to this sobering conclusion. "Alas, no matter how much they may wish it to be so, the proponents of alternatives—and better yet, 'clean' energy—cannot overcome the problem of scale," says Robert Bryce, a senior fellow at the Manhattan Institute. "Indeed, had any of the myriad advocates for renewable energy bothered to use a simple calculator, they would see that their favored sources simply cannot provide the vast scale of energy needed by the world's 7 billion inhabitants, at a price that can be afforded."[22]

The infrastructure for supplying the world's energy needs is also vast and complex. Replacing it with one based on renewable sources would be a massive undertaking. According to Vaclav Smil, a leading expert on energy and power systems, "[The current energy infrastructure] constitutes the costliest and most extensive set of installations, networks, and machines that the world has ever built, one that has taken generations and tens of trillions of dollars to put in place. . . . It is impossible to displace this supersystem in a decade or two—or five, for that matter."[23]

As politicians awaken to these realities, support for renewable energy is beginning to fade. According to data from the Bloomberg business website, US investment in renewable energy and energy efficiency dropped 54 percent in 2012.

> "[The current energy infrastructure] constitutes the costliest and most extensive set of installations, networks, and machines that the world has ever built, one that has taken generations and tens of trillions of dollars to put in place. . . . It is impossible to displace this supersystem in a decade or two—or five, for that matter."[23]
>
> —Vaclav Smil, a leading expert on energy and power systems.

Problems with Land Capacity

For renewable energy to replace fossil fuels for power generation—or even make a sizable dent in their use—an enormous amount of land is required. The installed arrays of solar panels or wind turbines would have to be ridiculously vast. Blogger Scott C. Johnston gave the following example: A typical coal-fired power plant is able to generate about 750 megawatts of electricity. A solar array the size of five hundred football fields would produce about 150 megawatts. So to fully replace one power plant, about 5 square miles (13 sq. km) of solar panels are

Intermittent Energy Output from Solar and Wind Sources

This graph shows solar and wind power generation from midnight to 6:00 p.m. for the California electrical grid on August 9, 2012. At 5:00 p.m., when energy demand peaked, solar and wind power production were far below their daily highs, showing how impractical alternative energy sources can be.

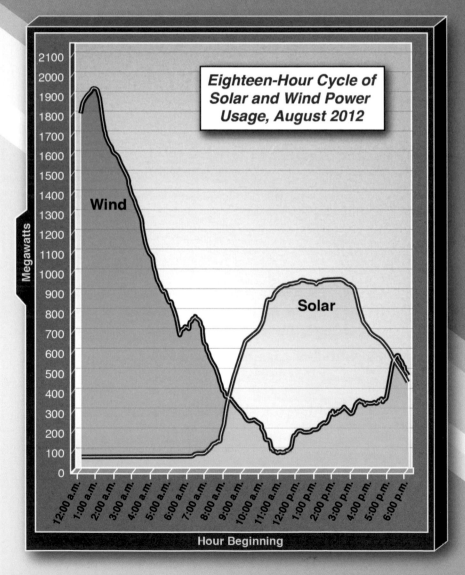

Eighteen-Hour Cycle of Solar and Wind Power Usage, August 2012

Megawatts

Wind

Solar

Hour Beginning

Source: Institute for Energy Research, "California's Flex Alert: A Case Study in Intermittent Energy," August 13, 2012. www.instituteforenergyresearch.org.

needed. A facility of that size is not easy to situate. As Johnston claims, "Supplanting our entire electrical supply with solar would require turning the entire state of South Carolina into one large solar panel."[24]

Some companies have used federal funds allocated for renewable energy projects to begin large solar plant projects in the Mojave Desert, where the sun shines three hundred days a year. But the large size of these projects worry environmental groups. Building and maintaining such facilities causes widespread disruption to desert wildlife. Under the federal Endangered Species Act, energy companies must estimate how many animals (individual animals, not species) would be displaced by an installation and also plan how to deal with this displacement. For all their concern about climate change and replacing fossil fuels, environmentalists are torn. "We've been supportive of efforts to accelerate clean-energy projects," said Jim Lyons, an expert on renewable energy at Defenders of Wildlife, a nonprofit conservation group. "But the scale of these new projects is massive, and they could be enormously destructive to plants and animals."[25]

Similar land-use problems apply to wind technology. To keep up with the growing global demand for electricity using wind power alone, the world's wind industry would have to cover 35,000 square miles (90,650 sq. km) with wind turbines—an area about the size of Indiana. Even if this could be done, where would all those turbines be placed? Wind currents atop mountains are best for power generation, but how many people want to see mountaintops fringed with wind turbines?

Intermittent Sources

Solar panels and wind turbines are only intermittent producers of electricity. If the sun is behind the clouds or has set, if the wind is not blowing, these energy sources are useless. The result is a blackout—a temporary but total loss of power—unless backup fossil fuel sources are available. Intermittency issues involve not only day-night variations, but also seasonal changes. Wind speeds tend to be higher at night than in the daytime, and also higher in winter than summer. Turbines cannot produce electricity at very low wind speeds; during storms with high winds, turbines must be turned off to avoid damage. Some experts suggest that the intermittency of wind power can be partially solved by spreading the turbines out over a larger area, thus taking advantage of different wind currents. However, this raises questions of efficient land use. As for solar energy, reduced sunlight produces less power in the winter, and of course sunlight is unavailable at night. Power grids that rely on sun and wind power to a significant degree must be upgraded to adapt to these fluctuations, or even redesigned at great cost. To avoid blackouts or brownouts in the meantime, coal and natural gas must generally be used for backup power. This switching back and forth can lead to other problems: For example, coal-burning facilities are baseload plants designed to run continuously and may falter or wear out if called upon to alternate with renewable power sources.

Lack of Storage

Renewable energy will never break through as a major supplier of electricity until the industry solves a major problem: finding a practical way to store energy. If not used at once, the energy produced by solar panels or wind turbines simply goes away. Thus, fossil fuels have a huge practical advantage over renewable sources. As Gregg Maryniak, chair of the Energy and Environmental Systems Track of Singularity University, puts it: "The phenomenon of the world's so-called addiction to fossil fuels is actually an aspect of a greater underlying energy truth. What society really wants and needs is *energy on demand*."[26] And for now and in the foreseeable future, renewable sources cannot provide it.

As with many renewable energy drawbacks, the problem has to do with scale. Portable energy storage in rechargeable batteries is fine for small-scale use in homes and businesses. There have even been improvements in battery performance. But currently the only way to store electricity on a large scale is a system called pumped hydroelectric energy storage. Water is pumped up to high-elevation reservoirs in off-peak hours and allowed to flow downhill through turbines when energy demand is high. This technique captures only about 70 percent of the stored energy and can only be used in certain geographical areas.

Environmental websites are full of ideas for energy storage, but they all tend to be expensive, impractical, or both. For example, two California companies, SolarReserve and BrightSource, are building plants that will use a technology called solar thermal to store power. A solar thermal plant uses the sun's heat to boil water, which in turn heats salt to store energy for use at night. Salt is used because it can store more heat than water. However, the salt must be kept molten or it will freeze to a solid that is difficult to remelt. The US Department of Energy has given SolarReserve a $737 million loan guarantee for a solar thermal plant that will store enough heated salt at one time to run for eight to ten hours—not much bang for the buck.

> "The phenomenon of the world's so-called addiction to fossil fuels is actually an aspect of a greater underlying energy truth. What society really wants and needs is *energy on demand*."[26]
>
> —Gregg Maryniak, chair of the Energy and Environmental Systems Track of Singularity University.

Of course, fossil fuels store energy all by themselves. The energy actually comes from sunlight and the process of photosynthesis. The sun's energy was stored as carbon in ancient plants that became fossilized. Each time a gasoline-powered car starts or a room lights up with electricity made from natural gas, there is a release of energy that has been stored—quite naturally—for more than 1 million years.

Chapter Three

Is Renewable Energy Too Expensive?

Renewable Energy Is Too Expensive

- Government subsidies paid for by higher taxes are required to fund renewable energy sources such as solar power and wind power.
- Producing energy from renewable sources is still much more expensive than fossil fuel sources.
- Building new connections from renewable energy sources to the main power grid requires an enormous capital expenditure, making those sources more expensive.

The Debate at a Glance

Renewable Energy Is Affordable

- Costs for producing energy from renewable sources are falling steadily.
- Flexible policies that encourage use of renewable energy also make it more cost-effective.
- Fossil fuels have hidden costs due to environmental damage and health risks that actually make renewable energy more competitive.

Renewable Energy Is Too Expensive

"There is growing evidence that the costs may be too high—that the price tag for purchasing renewable energy, and for building new transmission lines to deliver it, may not only outweigh any environmental benefits but may also be detrimental to the economy, costing jobs rather than adding them."

—Robert Bryce, a senior fellow at the Manhattan Institute.

Robert Bryce, "The High Cost of Renewable-Electricity Mandates," Manhattan Institute for Policy Research, February 2012. www.manhattan-institute.org.

Consider these questions as you read:

1. How persuasive do you find the arguments that renewable energy is too expensive? Explain.
2. Environmentalists contend that the ecological benefits of renewable energy should be factored into their cost. Do you agree? Why or why not?
3. Do you think consumers would be willing to pay more for electricity in order to adapt the power grid to renewable energy sources? Explain.

Editor's note: The discussion that follows presents common arguments made in support of this perspective, reinforced by facts, quotes, and examples taken from various sources.

Renewable energy is too expensive. Every year there are new predictions that solar, wind, biomass, or some other renewable technology will soon reach cost parity with coal, oil, and natural gas. However, this parity almost always depends on some sort of subsidy or government price support. Renewable energy cannot compete with fossil fuels without government subsidies. The reason that equipment costs for solar and wind power have fallen so much is that subsidies encouraged manufacturers to overproduce. For example, subsidies to solar firms from the Chinese and German governments led to a glut of solar panels worldwide and plummeting prices in the last decade. With those subsidies now drying up, it is questionable whether prices for solar panels will continue the downward trend.

In 2011 about two-thirds of the $24 billion that the United States spent on energy subsidies went to renewable sources such as solar, wind, and biofuels. Now, however, those kinds of subsidies are in jeopardy. According to a study by the Brookings Institution and two other energy-based think tanks, federal programs that support renewable energy are likely to expire in 2014. The Brookings Institution's Mark Muro, one of the report's authors, contends, "Neither Congress' attitude or the public purse right now is going to allow business as usual."[27]

> "When [alternative energy] subsidies and tax breaks go away, research and development will stagnate, prices will not get lower and the technology will not improve. The energy consumer will no longer have an incentive to choose clean alternative energy over fossil fuels."[28]
>
> —Environmental blogger Jeff Danovich.

Environmental blogger Jeff Danovich is more emphatic: "When these subsidies and tax breaks go away, research and development will stagnate, prices will not get lower and the technology will not improve. The energy consumer will no longer have an incentive to choose clean alternative energy over fossil fuels."[28]

All this comes after a spending spree that made the United States the world's number one supporter of renewable power. The anxiety expressed by those who favor renewables shows that they question whether these industries can make it on their own. The Brookings report recommends tripling government funding of research and development of renewable energy as well as increasing loan guarantees to help private firms market their products more quickly. However, with the failure of renewable energy firms like Solyndra still in the public's mind, it is doubtful that another round of

Increases in Germany's Electricity Prices to Support Renewable Energy

As a result of taxes to support renewable energy sources, Germany's price of electricity per kilowatt hour has steadily Increased. A 2012 initiative led to a 47 percent increase in one year.

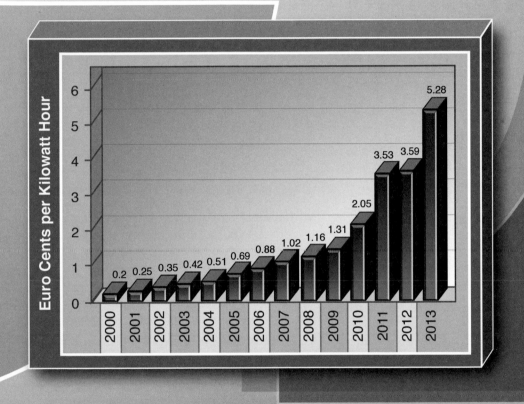

Source: *The Energy Collective,* "Renewable Energy and the Law of Receding Horizons," May 28, 2013. www.theenergycollective.com.

generous subsidies is on the way. As some critics have suggested, the "green" in "green energy" is mostly wasted money. And with the tepid global economy of the last few years, the massive financial outlays needed to support renewable energy is not affordable—by either public or private sources.

Higher Costs

Consumers are beginning to understand the economic reality of renewable energy: It costs more than energy from fossil fuels. As this sinks in, even many devoted supporters of renewable power become skeptics. Germany is a good example. In 2011 German chancellor Angela Merkel announced a major new energy plan: Prompted by concern about a nuclear accident in Japan, Germany would phase out its nuclear power plants by 2022 and concentrate on renewable energy. Merkel aimed for renewable energy to supply 80 percent of German power generation by 2050. To accomplish this the German government set up energy taxes to guarantee above-market prices for solar, wind, and biomass power. Polls at the time showed that many German citizens favored the plan. Energy experts hailed the decision as likely to lead to more efficient power generation with renewables.

However, today that enthusiasm has soured. The plan has made power for German households the most expensive in Europe—which was already expensive compared with that of the United States. German industry has been hit even harder. According to the Bloomberg business website, Germany's chemical industry alone paid 800 million euros in government fees to support renewables in 2013 after paying 550 million euros in 2012. (One euro is about 1.34 US dollars.) Some firms announced plans to move out of Germany to avoid the levies. Environment minister Peter Altmaier proposed freezing the extra charges to help consumers and businesses. "It's not justifiable that private electricity users, Mittelstand [mid-level industries] and small tradesmen carry alone the risk of electricity price increases as a result of the energy transition," Altmaier said. "We have now reached the limit."[29] In June 2013 Merkel reversed course and announced her intention to cut back on the renewable energy program.

> "It's not justifiable that private electricity users, Mittelstand [mid-level industries] and small tradesmen carry alone the risk of electricity price increases as a result of the energy transition. We have now reached the limit."[29]
>
> —Peter Altmaier, German environment minister.

The German magazine *Der Spiegel* said no one could have foreseen the cost explosion in renewable energy. Germany's experience is a cautionary tale for other countries such as the United States that are pushing for increased use of renewable energy.

Spain provides another sobering tale. Its aggressive approach to subsidizing renewable energy enabled it to receive 54 percent of its electricity from renewable sources. Utility companies had to pay large sums to renewable energy generators, yet they could not recoup their costs. This policy kept user rates low as supplier costs skyrocketed, so the real costs were never passed on to the consumer. Eventually the utility companies' deficit reached $7.3 billion in 2012. Spanish utilities now plan to issue bonds to pay for the deficit, despite a credit downgrade from two of the world's major rating companies, both of whom predict more deficits in years to come. One likely solution is that Spanish utility customers will end up paying much higher rates for electricity. The whole episode demonstrates the danger when the true costs of renewable energy are concealed from consumers.

The Expense of Adapting the Grid

Another concern associated with renewable energy is a reliable power grid. A power grid is the national network that carries electricity from generation plants to consumers, and it consists of wires, substations, switches, and transformers. Large power plants are based on the idea of constant generation and predictable capacity. Grid operators must constantly work to match power plant production to the changing needs of customers. Disruptions come when unpredictable sources of electricity, like solar or wind, are added to the grid. Too much input during a sunny or windy day can cause power surges; sudden drops in supply due to a thunderstorm can lead to troublesome brownouts. Corrections to output have to be immediate. And it turns out that adapting the grid to the fluctuations of renewable sources is very expensive.

Environmental engineers hope the solution is a smart grid that connects electric-powered devices just as the Internet connects computers. The idea is that as the cost of wireless communication devices comes down, the large number of sensors required all along the power grid can be installed affordably. So far, however, the evidence that such a system can be developed and implemented is not encouraging. For one thing, as more renewable sources are connected to the grid, expensive backup generators also have to be added and synchronized to the renewable energy supply. Of course, the backup generators run on nonfluctuating fossil fuels such as natural gas or oil. Paying for the smart grid usually requires bond issues, tax money, and federal assistance. As with most large infrastructure projects, costs can soar. For example, Chattanooga, Tennessee's, price tag for a smart grid upgrade has risen from an original $226 million to an estimated $552 million. That works out to $3,266 per customer—not exactly a negligible sum. As with so many aspects of renewable energy, the actual expense outweighs the theoretical benefit.

Renewable Energy Is Affordable

"As their costs continue to fall, renewable power sources are increasingly standing on their own merits versus new fossil-fuel generation."

—Maria van der Hoeven, International Energy Agency.

Quoted in Silvio Marcacci, "IEA: Renewables to Exceed Natural Gas, Nuclear Energy by 2016," *The Energy Collective* (blog), June 28, 2013. http://theenergycollective.com.

Consider these questions as you read:

1. Do you agree that renewable energy can be competitive with fossil fuels without the help of price supports or other government policies? Why or why not?
2. How important to consumers are the hidden costs associated with fossil fuels? Explain.
3. When you consider the facts and assertions in this discussion, do you agree with the argument that renewable energy is affordable? Which facts and ideas do you consider most persuasive?

Editor's note: The discussion that follows presents common arguments made in support of this perspective, reinforced by facts, quotes, and examples taken from various sources.

For years environmentalists claimed that renewable energy solutions were so important that cost did not matter. Some even claimed that fossil fuel prices *should* soar to discourage energy use and make renewable energy more competitive. During his 2008 presidential campaign, Barack Obama told the *San Francisco Chronicle*, "Under my plan of a cap-and-trade system, electricity rates would necessarily skyrocket."[30] His future energy secretary, Steven Chu, said in an interview, "Somehow, we have to figure out how to boost the price of gasoline to the levels in Europe"[31]— which was about eight to ten dollars a gallon. However, recent Gallup polls have shown that Americans prefer economic growth over environmental protection—a change from thirty years of opposite opinions. This emphasis means that renewable energy must be cost-effective to maintain support. And indeed, as technologies have improved and investment has continued to grow, the prices for renewable energy have fallen dramatically. This bodes well for the renewables industry since consumers no longer face such a stark choice between clean energy and cheap energy.

Falling Costs of Renewable Energy

The economic future of renewable energy can be stated in one word: *competitiveness*. "Future growth of renewable energy will increasingly be driven by cost competitiveness with fossil-fuel based generation,"[32] said Ivan Marten, senior partner at Boston Consulting Group. Experts have long claimed that renewable energy costs must fall before these sources can compete with fossil fuels. And that is exactly what is happening. According to the environmental website CleanTechnica, the prices for solar PV modules have dropped 80 percent since 2008, and the prices for wind turbines have fallen 29 percent. While there are many reasons for these price declines, including overproduction and reduced costs for raw materials, the overall effect is a much brighter future for renewables. As a 2012 study by Citibank observed, "Renewables will reach cost parity with conventional fuels (including gas) in many parts of the world in the very near term."[33]

> "Future growth of renewable energy will increasingly be driven by cost competitiveness with fossil-fuel based generation."[32]
>
> —Ivan Marten, senior partner at Boston Consulting Group.

For now, falling prices are often a hardship for suppliers in the industry. More than two dozen solar and wind manufacturers have gone bankrupt, and the surviving companies are feeling the pain. Yet lower prices for solar panels and wind turbines are making new projects much more attractive to developers. The day when renewable energy can compete with oil and coal on overall cost seems to be fast approaching, and investors are taking notice. Forecasters see investment in renewable sources tripling by 2030. "The apocalyptic views about what it will cost to shift the world to renewable energy simply aren't true," claimed Michael Liebreich, chief executive of Bloomberg New Energy Finance. "Three years ago [in 2010], we thought wind and solar would be cheap as chips, and they've even gone below that."[34] In other words, renewable energy is not only good for the environment, it is also proving to be good business.

Flexible Policies

Renewable energy also becomes more affordable when state and local governments find new ways to support the technology. Homeowners who install solar panels to generate power can connect their systems to the grid or use them as stand-alone (off-grid) installations for personal use. If connected to the grid, a homeowner can sell any excess power produced back to the local utility, thus reducing overall costs. Local regulations about land use, building codes, aesthetics, and noise control can affect a homeowner's ability to use solar panels or wind turbines. A flexible approach to these property laws can encourage more use of renewable energy. State and local governments can also focus more on standardizing grid-connection requirements to make it easier for homeowners to adapt.

Environmentally conscious corporations can also benefit from flexible policies. For example, a large company and large energy buyer like Walmart can promote lower costs for renewable energy. The company has secured several long-term agreements to purchase power from renewable sources and is seeking more. When a large corporation makes such a commitment, it can get low-cost financing that results in renewable energy delivered at or below fossil fuel prices. Miranda Ballentine, director of sustainability for Walmart, has studied the challenge and possible rewards of using renewable energy. "So, what's stopping Wal-Mart and other big energy buyers from doing more of this?" she asked. "You guessed it: policies."[35] Ballentine sees the need for faster permitting procedures with less red tape, allowing companies to make their own deals to get power from renewable energy suppliers at reasonable prices. As more companies pursue these agreements, renewable energy costs will continue to fall.

Hidden Cost of Fossil Fuels

While the cost of renewable energy is falling, the hidden cost of fossil fuels continues to rise. Debates comparing the cost of renewable energy to that of oil, natural gas, and coal often ignore these hidden costs (also called externalized costs). They include environmental impacts such as global warming and water and air pollution and health impacts such as lung disease. For example, many people assume coal combustion is much cheaper for producing electricity than solar plants or wind farms. In 2011 coal combustion produced about 42 percent of all electricity in the United States, and the consumer cost of coal-fired power remains quite low compared with other energy sources. Nevertheless, mining and burning coal is responsible for all sorts of negative effects. Strip mining for coal deposits disturbs the surface of the land and often pollutes waterways and kills fish. In mountaintop removal, forests are clear-cut to expose the top of a mountain, which is then blown off with explosives and mined with heavy equipment. This technique has altered the landscape in the Appalachian Mountains and clogged streams with rubble and waste. Coal mining companies are required to restore mined land to its original condition, but restoring ecosystems and habitats is a difficult procedure, and often the companies' efforts fall far short of this goal. Cleanup of a 2009 coal ash spill at a Tennessee Valley Authority power plant cost $825 million, much of it financed by rate hikes for customers.

> "The true cost of energy includes not just the price we pay at the gas pump or what shows up on the electric bill, but also the less obvious impact of energy use on health, the environment, and national security."[36]
>
> Michael Greenstone and Adam Looney, authors of a Hamilton Project report on the true costs of energy.

Using coal as an energy source is even more detrimental. According to the EIA, burning coal produces several emissions that are harmful to the environment and peoples' health, including sulfur dioxide, nitrogen oxides, carbon dioxide, various particulates, and fly ash residue. Joint efforts between the federal government and the coal industry to develop technologies for cleaner-burning coal have had some success but not enough to

The Falling Price of Solar Power

The price of solar power has plummeted since 1980, making it competitive with the prices of conventional energy such as crude oil, natural gas, and residential electricity (which includes power from a variety of sources).

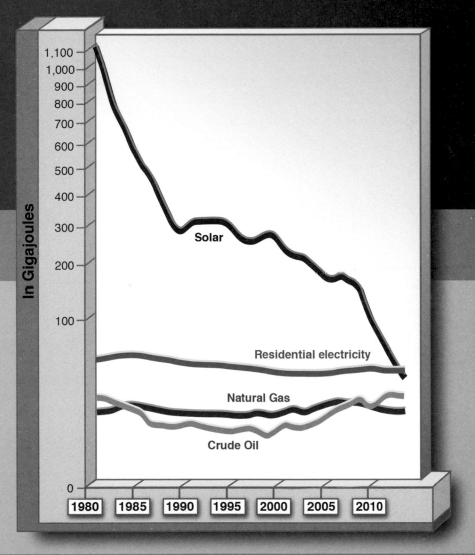

Note: A gigajoule is 1 billion joules of energy, or about 278 kilowatt hours.

Source: Thinkprogress, "Four Must-See Charts Show Why Renewable Energy Is Disruptive—in a Good Way," May 6, 2013. http://thinkprogress.org.

offset the drawbacks. The Clean Air Task Force contends that pollution from coal-fired plants causes 20,000 heart attacks, more than 9,000 hospitalizations, and more than 200,000 asthma attacks each year. A report by the Hamilton Project in May 2011 estimated the true externalized cost of coal-produced energy to be almost twice the retail price. With this in mind, renewable energy becomes much more competitive. As the authors of the Hamilton Project report state, "The true cost of energy includes not just the price we pay at the gas pump or what shows up on the electric bill, but also the less obvious impact of energy use on health, the environment, and national security. Economists refer to this more holistic accounting as the 'social costs' of energy consumption."[36] Environmental groups must continue to educate the public about the actual cost of fossil fuels so that the true value of renewable energy is better understood.

Chapter Four

Should the Government Help Develop Renewable Energy?

The Government Should Help Develop Renewable Energy

- The government has always played an important role in developing energy sources.
- Government support of renewable energy can lead to many new green jobs.
- Government-sponsored research can be crucial in developing renewable energy.
- Government incentives via the tax code can spur the growth of the renewable energy industry.

The Debate at a Glance

The Government Should Not Help Develop Renewable Energy

- The government should let the free market determine which, if any, renewable energy sources are practical and efficient replacements for fossil fuels.
- Government subsidies and regulations tend to distort the energy market and make all fuel costs higher.
- The failure of companies such as Solyndra show how risky and foolish it is for the government to try to pick winners and losers in the rapidly changing renewable energy industry.

The Government Should Help Develop Renewable Energy

"The government has always played a role in energy generation, and it will continue to do so. The question is what kind of energy should our leaders focus on: A mature fossil fuel industry that relies on dwindling resources and endangers our families with toxic pollution? Or the clean energy sector that taps the inexhaustible power of wind and solar and reasserts America as a technological powerhouse?"

—Peter Lehner, executive director of the Natural Resources Defense Council.

Peter Lehner, "Spur Innovation Not Last Century's Fuels," *Energy Experts Blog, National Journal*, March 15, 2012. http://energy.nationaljournal.com.

Consider these questions as you read:

1. Which arguments in favor of government support for renewable energy do you find most persuasive? Why?
2. Is developing infrastructure for using renewable energy a task that is well suited to the federal government? Why or why not?
3. Renewable energy industries receive much of their government support through the tax code. What are the advantages of this approach?

Editor's note: The discussion that follows presents common arguments made in support of this perspective, reinforced by facts, quotes, and examples taken from various sources.

The government should definitely help develop renewable energy as a replacement for fossil fuels. Those who think otherwise ignore the fact that for more than a century the government has consistently invested in energy to diversify the nation's supply and boost the economy. In the 1800s the federal government made cheap land grants to coal and timber interests. To boost petroleum industries, the government granted access to resources on public lands, subsidized exploration and drilling, and helped build railroads, waterways, and highways to transport the product. With the advent of electric power, Congress created the Federal Power Commission to hasten the building of dams and hydroelectric power plants to supply electricity. "Would the dams at Niagara Falls or Grand Coulee have been built by a group of ambitious entrepreneurs and a handful of angel investors?" asks John Plaza of Seattle-based Imperium Renewables. "Of course not. It took a combination of entrepreneurial spirit, back-breaking labor, long term vision and investments by the federal government."[37]

Congress also invested in the necessary infrastructure to connect rural areas to the power grid. The government sponsored important research into nuclear energy and then continued its support with loan guarantees and tax incentives for the nuclear industry. Congress long ago rewarded fossil fuel industries with subsidies that they still enjoy. It would be a reversal of long-standing policy if the government *did not* support promising new energy technologies such as solar, wind, geothermal, and biomass—especially since energy independence and a clean environment are so important to the national interest today.

Government support for renewable energy is growing at the state and local level as well. Currently twenty-nine states and the District of Columbia have passed renewable energy standards. These require power companies to get at least part of their electricity from wind, solar, or other renewables. As Brad Plumer, the *Washington Post*'s energy blogger, wrote, "Those state-level standards have played a big role in doubling the amount of renewable-energy capacity in the United States in the past four years."[38]

New Green Jobs

One of the chief benefits of government support for renewable energy is creating jobs—and more jobs than are provided by the oil industry. As the Union of Concerned Scientists observed, "Compared with fossil fuel technologies, which are typically mechanized and capital intensive, the renewable energy industry is more labor-intensive. This means that, on average, more jobs are created for each unit of electricity generated from renewable sources than from fossil fuels."[39]

At a time when unemployment figures have been troublesomely high, the prospect of long-term, well-paying jobs in renewable energy and related fields is welcome. Studies in the United States show there are about 100,000 people working full-time or part-time in the solar industry, 75,000 full-time employees in the wind industry, and 250,000 in hydroelectric power. Renewable energy holds opportunities for architects, engineers, designers, technicians, salespeople, and advertisers. In solar power alone there are jobs available in designing, manufacturing, installing, and selling solar panels and solar installations for homes and businesses, and similar positions exist in the wind, geothermal, and biomass industries. There are also agencies that specialize in matching applicants with renewable energy–related jobs, such as Solpower Jobs and Bright Green Talent. And these kinds of job opportunities are found not only at renewable energy companies. "Although 'green' hiring is often associated with employment at green companies," said James Beriker, the CEO of an employment website, "we are seeing more and more roles at traditional companies focused on sustainable practices as a strategic and competitive business advantage. As a result, green energy and technology fields continue to grow."[40] Government subsidies and tax policies help sustain this job growth going forward.

> "Although 'green' hiring is often associated with employment at green companies, we are seeing more and more roles at traditional companies focused on sustainable practices as a strategic and competitive business advantage. As a result, green energy and technology fields continue to grow."[40]
>
> —James Beriker, CEO of an employment website.

Government Research Is Crucial

Government support for renewable energy technologies should also include research and development. This means providing federal grants for research projects at universities and scientific institutes. Researchers and scientists can use their work to demonstrate the potential for different types of renewable energy to policy makers and companies. They can also propose innovations to industry or partner with business interests to develop promising ideas. Technologies such as supercomputing, GPS navigation, and the Internet all received a huge boost from government-sponsored research. "If you take any major information technology company today . . . you can trace the core technologies to the rich synergy between federally funded universities and industry research and development,"[41] said Peter Lee, a vice president of Microsoft Research.

Lee helped produce a 2012 report by the National Research Council that traced the development of eight computer technologies from government-paid research to commercial use. The report concluded that the eight government-sponsored projects ended up helping produce nearly $500 billion in corporate revenue. Not bad for seed money—and just what the growing renewable energy sector desperately needs. Up to now, government support for research into renewable energy technology has been erratic and highly politicized. This must change if renewable energy is to reach its potential in the United States.

Incentives for Growth

One of the government's most effective tools to support renewable energy is tax policy. Congress should maintain tax laws that encourage more investment in renewables by businesses and individuals. With tax policy, the government does not spend money directly on renewable energy. Instead, it supports renewable energy through reducing the tax bills of companies and consumers. This is done with tax credits, deductions, exclusions, exemptions, and preferential tax rates. Since each source of renewable energy has different characteristics and advantages, Congress can adapt tax laws to provide the most effective support for solar, wind, biomass, and other technologies. Two important tax-code tools are the production tax credit and the investment tax credit.

Probably the most crucial of these is the renewable electricity production tax credit, or PTC. Under the PTC the owner of a renewable

Average Annual Energy Subsidies from the US Government

The United States government has historically spent more annually on subsidies to oil and natural gas than it has on renewable energy sources. Subsidies are valuable in helping renewable industries become viable.

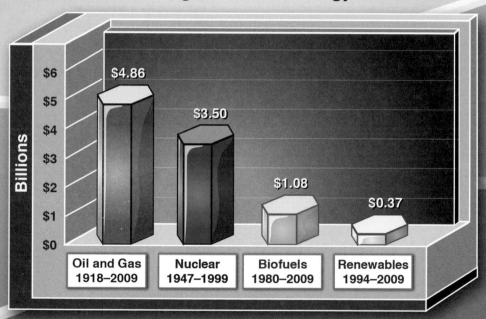

Historical Average of Annual Energy Subsidies

Billions

$4.86	Oil and Gas 1918–2009
$3.50	Nuclear 1947–1999
$1.08	Biofuels 1980–2009
$0.37	Renewables 1994–2009

energy project such as a wind farm gets a tax credit for each kWh of electricity produced. This means that the owner can reduce his or her year-end tax bill by a specified amount—which was 2.2 cents per kWh in 2012. For example, a wind farm that generates 260 million kWh's per year earns more than $5 million in PTC tax credits for its operator. As Richard W. Caperton, director of clean energy at the Center for American Progress, explains, "This tax credit is economically the same as government spending: The government has less money than it would have without the investment, and the project is more profitable. It is also true that the incentive helped stimulate the investment that made both the income and the tax expenditure possible. In short, this investment helped directly create economic activity and growth."[42] Since creating the PTC in 1993, the government has used it to invest billions of dollars in renewable technologies, mainly wind power. And the investment has been a success. In this period more than 40 gigawatts of new wind power capacity has come on-line. One gigawatt is enough energy to power 1 million 100-watt lightbulbs.

The investment tax credit gives re-newable energy project developers an up-front tax credit based on the amount of the initial investment. This works well with solar projects, which are expensive to install and slower to achieve profit-ability. As with the PTC, the investment tax credit has been effective in lowering overall costs for the solar industry, making solar power more competitive with fossil fuel energy. The 2009 stimulus bill extended the credit through 2016.

> "This tax credit is economically the same as government spending: The government has less money than it would have without the investment, and the project is more profitable. . . . In short, this investment helped directly create economic activity and growth."[42]
>
> —Richard W. Caperton, director of clean energy at the Center for American Progress.

Over the past decade, tax policies that boost renewable energy have increased worldwide. As of 2013 sixty-six countries supported renewable energy with their tax codes, resulting in a remarkable growth in the sec-tor. Renewable sources are beginning to take over a larger share of power generation, and polluting fossil fuel plants are being retired. Continued tax incentives are necessary so that this success story is not threatened.

The Government Should Not Help Develop Renewable Energy

"As the chief executive of a solar technology company, no one wants an abundant supply of clean energy and a healthy solar energy industry more than I do. And the best pathway to a stable renewable energy industry is to create self-sufficiency and independence from government financial assistance.

—Paul Nahi, CEO of Enphase Energy, a provider of systems for the solar industry.

Paul Nahi, "Government Subsidies: Silent Killer of Renewable Energy," *Forbes*, February 14, 2013. www .forbes.com.

Consider these questions as you read:

1. Is the free market the best way to determine the viability of a renewable energy technology? Explain.
2. How persuasive is the argument that government subsidies for renewable energy distort the market for electricity overall? Explain.
3. How can government support renewable energy without resorting to "picking winners and losers" among renewable energy companies?

Editor's note: The discussion that follows presents common arguments made in support of this perspective, reinforced by facts, quotes, and examples taken from various sources.

The free market should determine the winners and losers in renewable energy, not the federal government. Companies that make solar panels or wind turbines must compete for business just like any other companies, including those that extract and market fossil fuels. According to the National Center for Policy Analysis, "The principles of the free market . . . remind us that if a technology is viable the market will ensure its success, and if a technology is not commercially viable no amount of government support can lead to its success."[43]

Many experts are skeptical about the ability of renewable energy to compete with fossil fuels, particularly since new extraction technologies are promising an abundance of cheap fossil fuel energy in the near future. Craig Pirrong, energy markets director for the Global Energy Management Institute, expressed the viewpoint of the skeptics: "All renewables are cursed with fundamental problems that make their future stand-alone (i.e., unsubsidized) viability as anything but a marginal energy source highly questionable." Pirrong does admit that renewable energy might find a place in the economy's energy mix and believes the free market should determine that place. "Insofar as renewables have desirable environmental attributes . . . the preferable approach is to price these attributes and let the market choose the technologies that produce the best balance between environmental and non-environmental considerations."[44]

> "The principles of the free market . . . remind us that if a technology is viable the market will ensure its success, and if a technology is not commercially viable no amount of government support can lead to its success."[43]
>
> —National Center for Policy Analysis.

Currently the government's renewable energy policy has two parts: supporting basic research and providing funding for business operations through loan guarantees and subsidies. Funding research into renewables is a defensible function of government, since fledgling companies often cannot afford such investments on their own. But the second part of government energy policy is misguided and wasteful. It would be better left to the free market. Jim Nelson, CEO of Solar3D, described three steps in bringing a renewable energy technology to market: innovation or research, developing a product, and expansion. "One of the greatest strengths in America is innovation," Nelson said. "It is a long and rich tradition for the U.S. to lead the world in innovation. The government currently plays a key role in providing funds to many companies in the proof-of-concept stage, as well as to national labs and universities developing new technologies. Steps two and three should be left to private investors."[45]

Subsidies Distort the Market

Up to now, the renewable energy industry has benefited greatly from government subsidies. However, these subsidies do more harm than good, as they distort the market for energy and distract companies from focusing on ways to be more competitive in the marketplace. "Government subsidies to new wind farms have only made the industry less focused on reducing costs," said Patrick Jenevein, CEO of the Tang Energy Group. "In turn, the industry produces a product that isn't as efficient or cheap as it might be if we focused less on working the political system and more on research and development."[46] Jenevein also points out that after the 2009 stimulus bill subsidies, the wind power industry began to build wind farms in less windy—and less profitable—locations simply because the money was available.

> "Government subsidies to new wind farms have only made the industry less focused on reducing costs."[46]
>
> —Patrick Jenevein, CEO of the Tang Energy Group.

Subsidies have encouraged solar companies to expand too rapidly also. For example, almost half of the 7.2 gigawatts of solar capacity in the United States were installed in 2012, a frenzied growth spurt that saw even reputable companies cutting corners by using cheaper materials or fewer manufacturing controls. Many in the industry felt that solar energy had to prove its viability very soon to keep the subsidy money flowing. The result, as the *New York Times* reported, has been rushed and shoddy manufacturing leading to a quality crisis in the solar industry. Panels built to last twenty-five years are failing after only two. Coatings designed to protect the panels are disintegrating under real-world conditions, while other defects have caused fires. Tim Worstall, a senior fellow at the Adam Smith Institute in London, thinks this widespread problem shows how the industry, flush with subsidy cash, is scrambling to reach impractical production goals. "Shoddy production is an inevitable outcome of an industry expanding pell mell," said Worstall. "Which would be an indication that we're just trying to do this switch over to solar too fast. Which in itself would be an indication that we've set those subsidies too high."[47]

Subsidies for renewable energy take many forms, but they usually mean that the federal or state government imposes a small tax or electricity rate hike on consumers. The government then gives the proceeds to companies that appear to have the technological skill and business savvy to thrive. Too often, however, the subsidy funds go to companies that merely have skillful lobbyists who know how to work the legislature or Congress. (The fossil fuel industry receives subsidies, too, but it has a proven ability to be profitable and satisfy energy needs.) Environmentalists justify these subsidies to renewables as a necessary means to create a system of clean energy. But it remains to be seen whether voters are sufficiently alarmed about global warming and other environmental issues to continue supporting large subsidies for renewable energy sources that are still not competitive, in cost or capacity, with fossil fuels.

Difficulty of Picking Winners

It is foolish to think that the government can pick the most promising technologies and the best-run companies in any industry, including renewable energy. Certainly the US government's recent track record does not inspire confidence. In 2012 the Heritage Foundation reported that more than thirty government-backed green energy companies had either gone bankrupt or were financially troubled. The notorious failure of solar company Solyndra is only one of the stories. Abound Solar, a manufacturer of thin-film solar panels, received $400 million in loan guarantees from the US Department of Energy in 2007. Due to the plummeting price of solar panels, the company declared bankruptcy in 2012, with the Energy Department still on the hook for $40 million to $60 million. Nevada Geothermal Power got a $98 million Department of Energy–backed loan to construct a geothermal plant outside of Reno. Not long thereafter, company audits revealed losses and debt equal to the original loan amount. As Ashe Schow of *The Foundry* blog writes, "The government's picking winners and losers in the energy market has cost taxpayers billions of dollars, and the rate of failure, cronyism, and corruption at the companies receiving the subsidies is substantial."[48]

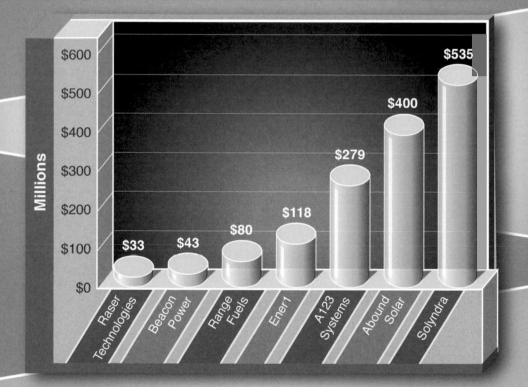

Government-Backed Renewable Energy Companies That Have Gone Bankrupt

Several renewable energy companies that received subsidies from the federal government have declared bankruptcy. Critics say this demonstrates the problems that arise when the government tries to pick winners and losers in industry. A sample of these companies appears here.

Millions

- Raser Technologies: $33
- Beacon Power: $43
- Range Fuels: $80
- Ener1: $118
- A123 Systems: $279
- Abound Solar: $400
- Solyndra: $535

Source: *The Foundry*, "President Obama's Taxpayer-Backed Green Energies Failures," October 18, 2012. http://blog.heritage.org.

Other governments are proving no better than the United States at picking renewable energy winners. In March 2013 China's Suntech Power, a heavily subsidized firm that once was the largest solar manufacturer in the world, declared bankruptcy. Analysts blamed its failure on oversupply of solar panels in the global market and trade disputes with the

United States and other countries. China also announced it was cutting back its aggressive Golden Sun program of subsidies for solar power. In China as elsewhere, renewable energy businesses must increasingly rely on the marketplace to survive.

With such a troubling record of failed investments and unmet goals in renewable energy, and with the added burdens of slow economic growth, governments worldwide are much less likely to wager large sums on choosing winners among renewable energy companies in the future. Thankfully for taxpayers, this wasteful procedure seems likely to end.

Source Notes

Overview: Renewable Energy

1. Quoted in Marc Gunther, "What Can We Learn from Solyndra's Failure?," *The Energy Collective* (blog), September 23, 2011. http://the energycollective.com.
2. James Hansen, "Why I Must Speak Out About Climate Change," TED talk (video), February 2012.

Chapter One: Are Renewable Energy Sources Needed?

3. Steven Koonin, "How Can We Transform the Energy Sector?," *World Economic Forum* (blog), March 6, 2013. http://forumblog.org.
4. John Upton, "Americans Want More Renewable Energy and More Climate-Change Prep," *Grist*, April 2, 2013. http://grist.org.
5. *American Energy Independence*, "America's Solar Energy Potential." www.americanenergyindependence.com.
6. Quoted in Michael Martina, "China Premier-Designate Says Pollution Solution 'Long-Term Process,'" Reuters, January 15, 2013. www.reuters.com.
7. Quoted in Andrew Freedman, "Fossil Fuels to Dominate World Energy Use Through 2040," Climate Central, July 25, 2013. www .climatecentral.org.
8. US Energy Information Administration, "Technically Recoverable Shale Oil and Shale Gas Resources: An Assessment of 137 Shale Formations in 41 Countries Outside the United States," June 10, 2013. www.eia.gov.
9. Quoted in Julia Borger and Larry Elliott, "How Cheap Energy from Shale Will Reshape America's Role in the World," *Guardian* (London), November 15, 2012. www.theguardian.com.
10. Quoted in Alice Baghdjian, "British Engineers Create Petrol from Air and Water," Reuters, October 19, 2012. www.reuters.com.

11. Quoted in Rachel Feltman, "Hijacking E. Coli to Brew Synthetic Fuel," *Popular Mechanics*, April 22, 2013. www.popularmechanics .com.

12. Quoted in News & Research Communications, "OSU Study Questions Cost-Effectiveness of Biofuels and Their Ability to Cut Fossil Fuel Use," Oregon State University, November 30, 2011. http:// oregonstate.edu.

13. Quoted in Jonathan Parkinson, "The Not-So-Sunny Side of Solar Panels," *Voice of San Diego*, February 16, 2009. http://voiceofsandiego .org.

Chapter Two: How Practical Is Renewable Energy?

14. Chris Williams, "Clean Energy Is Possible, Practical, and Essential— Now!," *Climate & Capitalism*, June 23, 2011. http://climateand capitalism.com.

15. Quoted in Mark Fischetti, "How to Power the World Without Fossil Fuels," *Scientific American*, April 15, 2013. www.scientificamerican .com.

16. Mark Halper, "Wind, Wind Everywhere. Who Needs Any Other Energy Source?," *Intelligent Energy* (blog), SmartPlanet, September 17, 2012. www.smartplanet.com.

17. Quoted in Unni Skoglund, "Geothermal Energy Could Provide All the Energy the World Will Ever Need," Renewable Energy World, September 16, 2010. www.renewableenergyworld.com.

18. Interview with the author, August 12, 2013.

19. Quoted in Nathanael Massey and ClimateWire, "Solution to Renewable Energy's Intermittency Problem: More Renewable Energy," *Scientific American*, December 12, 2012. www.scientificamerican.com.

20. Quoted in Silvio Marcacci, "Solar Energy Forecasts Could Boost Industry Outlook," CleanTechnica, February 28, 2013. http://clean technica.com.

21. Quoted in Ken Silverstein, "Obama Nourishes Smart Grid and Feeds His Legacy," *Forbes*, June 19, 2013. www.forbes.com.

22. Quoted in *Energy Tribune*, "Renewable Energy's Incurable Scale Problem," July 13, 2012. www.energytribune.com.

23. Vaclav Smil, "A Skeptic Looks at Alternative Energy," *IEEE Spectrum*, June 28, 2012. http://spectrum.ieee.org.

24. Scott C. Johnston, "The Idiot's Guide to Why Renewable Energy Is Not the Answer," *The Naked Dollar* (blog), June 1, 2010. http://the nakeddollar.blogspot.com.

25. Quoted in Kiera Butler, "Big Solar's Death Panels," *Mother Jones*, March/April 2011. www.motherjones.com.

26. Gregg Maryniak, "Storage, Not Generation, Is the Challenge to Renewable Energy," *The Blog, Huffington Post*, July 25, 2012. www.huff ingtonpost.com.

Chapter Three: Is Renewable Energy Too Expensive?

27. Quoted in Matt Smith, "U.S. Faces Clean Energy Bust as Subsidies Expire, Report Warns," CNN, April 18, 2012. www.cnn.com.

28. Jeff Danovich, "Green Energy Will Win with Government Subsidies and Tax Incentives," Policymic, 2012. www.policymic.com.

29. Quoted in Sheenagh Matthews, "German Green Energy Push Bites Hand That Feeds Economy," Bloomberg, January 28, 2013. www .bloomberg.com.

30. Quoted in PolitiFact, "Obama Said Cap and Trade Would Increase Electricity Rates," November 17, 2009. www.politifact.com.

31. Quoted in Neil King Jr. and Stephen Power, "Times Tough for Energy Overhaul," *Wall Street Journal*, December 12, 2008. http://on line.wsj.com.

32. Quoted in *Wall Street Journal*, "The Experts: What Renewable Energy Source Has the Most Promise?," April 17, 2013. http://online .wsj.com.

33. Citi Research, "Shale and Renewables: A Symbiotic Relationship," September 12, 2012. www.ourenergypolicy.org.

34. Quoted in Louise Downing and Alex Morales, "Renewables Investment Seen Tripling Amid Supply Glut," Bloomberg, April 21, 2013. www.bloomberg.com.

35. Miranda Ballentine, "The Secret to Affordable Renewable Energy," *World Economic Forum Blog*, November 30, 2012. http://forum blog.org.

36. Michael Greenstone and Adam Looney, "A Strategy for America's Energy Future: Illuminating Energy's Full Costs," Hamilton Project/ Brookings, May 2011. http://brookings.edu.

Chapter Four: Should the Government Help Develop Renewable Energy?

37. John Plaza, "The U.S. Government Has a Long History of Financing Energy Infrastructure," Renewable Energy World, October 19, 2009. www.renewableenergyworld.com.
38. Brad Plumer, "The Biggest Fight over Renewable Energy Is Now in the States," *Washington Post*, March 25, 2013. www.washingtonpost .com.
39. Union of Concerned Scientists, "Benefits of Renewable Energy Use: Jobs and Other Economic Benefits," April 8, 2013. www.ucs usa.org.
40. Quoted in Jacquelyn Smith, "The Top 10 Cities for Green Jobs," *Forbes*, May 7, 2013. www.forbes.com.
41. Quoted in Steve Lohr, "The Seeds That Federal Money Can Plant," *New York Times*, October 6, 2012. www.nytimes.com.
42. Richard W. Caperton, "Fair, Effective, and Efficient Tax Policy Is Key for Driving Renewable Energy Growth," Center for American Progress, January 10, 2012. www.americanprogress.org.
43. National Center for Policy Analysis, "Wind Energy Cannot Compete in Free Market," April 25, 2013. www.ncpa.org.
44. Quoted in *Wall Street Journal*, "The Experts."
45. Jim Nelson, "U.S. Government Should Trust the Free Market for Green Energy Investment," Renewable Energy World, May 29, 2012. www.renewableenergyworld.com.
46. Patrick Jenevein, "Wind-Power Subsidies? No Thanks," *Wall Street Journal*, April 1, 2013. http://online.wsj.com.
47. Tim Worstall, "Solar Power Subsidies Were Too Large Too Fast," *Forbes*, May 30, 2013. www.forbes.com.
48. Ashe Schow, "President Obama's Taxpayer-Backed Green Energy Failures," *The Foundry* (blog), Heritage Foundation, October 18, 2012. http://blog.heritage.org.

Renewable Energy Facts

Renewable Energy Use

- The International Energy Agency says that electricity generation from renewable energy sources was nearly 20 percent of the world's total in 2012. At its current rate of growth, renewable energy will soon surpass natural gas and become second only to coal as a global energy source.
- According to a 2013 report in the *New York Times*, Iceland generates 100 percent of its electricity from renewable energy sources, including hydroelectric and geothermal. Norway gets 97 percent of its electricity from renewable sources, and Canada gets 63 percent. By contrast, the United States gets about 13 percent of its electric power from renewable sources.
- The top five nations in percentage of the world's renewable energy use are the United States, China, Germany, Spain, and Brazil.
- The EIA reported in 2012 that the top five states for generating non-hydro renewable energy were Texas, California, Iowa, Minnesota, and Washington. Wind power made up most of the total in all of these states except California, which produced more geothermal power.

Most Used Sources of Renewable Energy

- According to the International Energy Agency, hydroelectric power is currently the world's most-used source of renewable energy by far. In 2011 it accounted for 16 percent of total power generation worldwide.
- The International Energy Agency also reported that in 2011 wind and solar power made up 4 percent of the world's total power generation. This share is predicted to double by 2018.
- In 2012 Reuters news service said that Germany had nearly as much installed capacity for solar energy generation as the rest of the world combined. Germany also has twenty-one times more solar installed per capita than does the United States.

- The US Department of Energy reported in 2012 that wind energy had become the top US source of new electricity generation for the first time. In the first four years of the Barack Obama administration, electricity generated from solar and wind more than doubled.

Personal Use of Renewable Energy

- The Mother Nature Network estimated that 750,000 Americans live "off the grid" by generating their own power from renewable energy sources.
- The Solar Energy Industries Association and GTM Research reported that in 2012 more than 16 million individual solar panels were installed in the United States, making solar power the fastest-growing domestic energy source. Almost one-third of those panels were installed in California.
- According to the Wind Energy Foundation, the purchase and installation of a wind turbine system large enough to power a home costs, on average, about $30,000, with a high-end cost of about $70,000. The federal government and many states have rebate and tax credit programs to support investment in home wind systems.
- In most US markets, the price of power from rooftop solar panels compares favorably with local retail prices for electricity. The fact that peak power demand, when electricity prices are highest, coincides with peak solar output makes solar energy competitive with or even cheaper than electricity from the grid.

Negative Factors

- In 2012 representatives of the United Kingdom's National Trust, a conservation charity, criticized wind farms as being unsightly and spoiling the landscape. Homeowners have also complained that the turbines are noisy and often keep them awake at night.
- A study by the Energy Institute at the Massachusetts Institute of Technology suggests that the electricity grid must be upgraded before intermittent renewable energy sources such as wind and solar can become a

significant factor in power generation. Such an upgrade will be expensive and technically challenging.

- In 2013 *USA Today* reported that Europe's focus on subsidizing renewable energy to bring it online more quickly has resulted in much higher prices for electricity. Critics argue that renewable energy sources must be subsidized to compete with fossil fuels, thus distorting the market for energy.

Positive Factors

- The Intergovernmental Panel on Climate Change reported that global warming emissions during the life cycle of various kinds of renewable energy were minimal. This life cycle includes manufacturing, installation, operation, and maintenance.
- Renewable energy can be a boon for the world's poorer nations. Kenyan families that once paid ten dollars a month for kerosene and two dollars a month for charging cell phones at a village center can now buy a solar set on an installment plan to generate their own power and save money.
- A research team at the University of Buffalo is developing a paint made of solar cells that could rapidly become cost competitive with ordinary house paint. The paint would enable homeowners to produce solar energy more cheaply than with solar panels.
- A study by the Reiner Lemoine Institut and Solarpraxis AG found that solar and wind power systems work much better together than previously understood. The study found that complementary solar photovoltaic systems and wind turbines provide twice the amount of generated electricity using the same surface area. One reason the systems work well together is because solar panels generate more power in the summer, while wind turbines are more productive in colder months.

Related Organizations and Websites

American Council on Renewable Energy (ACORE)
1600 K St. NW, Suite 650
Washington, DC 20006
phone: (202) 393-0001
e-mail: shays@acore.org
website: www.acore.org

ACORE is a nonprofit organization that wants to build a secure and prosperous America through clean, renewable energy. The group focuses its efforts on the economic, environmental, and security benefits of renewable energy. ACORE members include financial institutions, government leaders, and industry professionals.

American Petroleum Institute (API)
1220 L St. NW
Washington, DC 20005
phone: (202) 682-8000
website: www.api.org

The API represents the US oil and natural gas industry. Its five hundred corporate members come from all segments of the industry. The API's mission is to promote public policy that maintains a strong and viable oil and natural gas industry.

American Solar Energy Society
4760 Walnut St., Suite 106
Boulder, CO 80301
phone: (303) 443-3130 • fax: (303) 443-3212
e-mail: info@ases.org
website: www.ases.org

The American Solar Energy Society is a nonprofit group that works to increase the use of solar energy and energy efficiency in the United States. Its membership of solar professionals and advocates seeks to further education, research, and policy related to solar energy.

American Wind Energy Association (AWEA)

1501 M St. NW, Suite 1000

Washington, DC 20005

phone: (202) 383-2500 • fax: (202) 383-2505

website: www.awea.org

The AWEA is the national trade group for the US wind industry. The association promotes wind energy as a clean source of electricity for consumers.

The Brookings Institution

1775 Massachusetts Ave. NW

Washington, DC 20036

phone: (202) 797-6000

e-mail: communications@brookings.edu

website: www.brookings.edu

The Brookings Institution is a nonprofit public policy organization that does independent research and offers innovative, practical recommendations on vital issues of the day, including energy policy. Brookings is consistently rated one of the most influential think tanks in the United States.

Center for American Progress (CAP)

1333 H St. NW

Washington, DC 20005

phone: (202) 682-1611

website: www.americanprogress.org

CAP is an educational institute that promotes progressive ideas to improve the lives of Americans. It is a supporter of government involvement in renewable energy projects.

Geothermal Energy Association (GEA)
209 Pennsylvania Ave. SE
Washington, DC 20003
phone: (202) 454-5261
website: www.geo-energy.org

The GEA is a trade group made up of US companies that are working to expand use of geothermal energy for electrical power generation. The GEA promotes policies that will develop geothermal resources and also engages in education and community outreach.

Intergovernmental Renewable Energy Organization (IREO)
884 Second Ave.
Dag Hammarskjöld UN Centre No. 20050
New York, NY 10017
phone: (212) 647-7000 • fax: (212) 202-4100
website: www.ireoigo.org

Formed in 2008 by member states of the United Nations, the IREO promotes the transition to sustainable development and renewable energy resources. The IREO supports collaboration between governments and the private sector for clean energy projects.

Sierra Club
85 Second St., 2nd Floor
San Francisco, CA 94105
phone: (415) 977-5500 • fax: (415) 977-5797
e-mail: information@sierraclub.org
website: www.sierraclub.org

The Sierra Club is an environmental group that promotes the responsible use of the earth's ecosystems and resources. It supports leaders and politicians who will work to expand clean energy markets and stop climate change.

For Further Research

Books

Godfrey Boyle, *Renewable Energy: Power for a Sustainable Future*. New York: Oxford University Press, 2012.

Michael Levi, *The Power Surge: Energy, Opportunity, and the Battle for America's Future*. New York: Oxford University Press, 2013.

Preben Maegaard, Anna Krenz, and Wolfgang Palz, eds., *Wind Power for the World: The Rise of Modern Wind Energy*. Boca Raton, FL: CRC, 2013.

John Perlin, *Let It Shine: The 6,000-Year Story of Solar Energy*. Novato, CA: New World Library, 2013.

Daniel Yergin, *The Quest: Energy, Security, and the Remaking of the Modern World*. New York: Penguin, 2012.

Ozzie Zehner, *Green Illusions: The Dirty Secrets of Clean Energy and the Future of Environmentalism*. Lincoln and London: University of Nebraska Press, 2012.

Periodicals

Diane Cardwell, "On Rooftops, a Rival for Utilities," *New York Times*, July 26, 2013.

Amory B. Lovins, "It Doesn't Matter If We Never Run Out of Oil: We Won't Want to Burn It Anymore," *Atlantic*, May 13, 2013.

Brad Plumer, "State Renewable-Energy Laws Turn Out to Be Incredibly Hard to Repeal," *Washington Post*, August 8, 2013.

Greg Pollowitz, "*NYT* Discovers 'Green' Power's Intermittency Problem," *National Review*, August 15, 2013.

Larry Sherwood, "Explosive Growth," *Solar Today*, July/August, 2013.

Wall Street Journal, "The Experts: What Renewable Energy Source Has the Most Promise?," April 17, 2013.

Internet Sources

CleanTechnica, "Shhhh! New Low Noise Wind Turbine Blades Designed by GE," August 19, 2013. http://cleantechnica.com.

Renewable Energy World, "Doing Good by Doing Solar," July 12, 2013. www.renewableenergyworld.com.

The Energy Collective (blog), "How Low Can Solar Energy Go? Empa's New Thin Film Breakthrough," August 17, 2013. http://theenergy collective.com.

Index

Note: Boldface page numbers indicate illustrations.

agriculture
    dependence of, 20, 21, 64
    wages and costs to consumers, 15, 20
Alabama, 59, 64
al Qaeda, 29, 30, 31
American Conservative (magazine), 49
American Immigration Council, 38
amnesty
    cost of public services and benefits, 50
    examples of other countries, 49
    past programs, 9, 40, 48
    real reason for granting, is political, 47–48
    rewards illegal behavior, 40
    should be offered, 40, 42
    See also citizenship
Amnesty Act (1986), 48
Antle, W. James, III, 49
Antonovich, Michael, 17
Arévalo Pedroza, Guillermo, 57
Arizona
    civilian border patrols in, 55–56
    cost to, 60
    crime in, 25, 28, 33, 35–36, 63
    fence along border with Mexico, 49–50
    "show me your papers" provision of SB 1070, 35–37
Arreguin, Antonio Medina, 28
Arvak, Jason, 36–37

Bargo, Michael, Jr., 61, 64
Beason, Scott, 59
Belien, Paul, 49
Bell, Tony, 17
Bermudez, Roy, 36
Bernacke, Michael, 49–50
Bloomberg, Michael, 43–44
Border Patrol
    cannot be effective, 44, 57, 58, 66
    civilian border patrols could assist, 55–56
    is effective, 40, 49–50
    members of, 58
    state and local law enforcement could assist, 55
    uses unnecessary lethal force, 57–58
    workplace sweeps by, 51
Boston marathon bombing, 30–31
Brinson, Randy, 59
Brussels Journal, 49
Bush, George W., 44, 49, 66
Bush, Jeb, 48

California
    cost to, 17, 18, 60
    crime in, 26, 27
    most wages are spent in, 22–23
    number in, 27
Center for American Progress, 24, 44
children
    benefits granted to US-born, 17
    of unauthorized immigrants
        deportation of, 43
        deserve citizenship, 43
        proposals for citizenship for those brought as, 9, 43
        residency path for, 9
        will become taxpayers, 23
    use of drugs by, 27–28
Chronicle (online magazine), 26
citizenship
    is morally wrong, 51
    is privilege to be earned, 52
    many unauthorized immigrants do not want, 48
    most Americans favor, 40, 42–43
    politics of, 47–48
    provides protections, 43
    rewards illegal behavior, 47, 51–52
    should be offered, 40, 41–44, 45, 46
    should not be offered, 40, 47–52, 50
    for those brought as children, 9, 43
civilian border patrols, 55–56
Colorado Alliance for Immigration Reform, 18
Constitution, 62, 66
construction industry, 20, **21**

Cornyn, John, 30
costs
    to construction industry, 20
    of food are lowered, 15, 20
    to taxpayers
        of amnesty, 50
        in Arizona, 60
        of Border Patrol, 44
        in California, 17, 18, 60
        of combating crime, 25, 36–37
        cost of aging population to, 12, 23,
            24
        of deportation, 43–44
        for education, 16, 18
        of federal enforcement, 65
        local, 59–60
        of public services, 12, 16
        in Texas, 60
crime/criminals
    cost of combating, 25, 36–37
    by country of origin, 29
    frequency of arrests of same individual,
        27
    incarceration rate, 27
    is major issue, 25
    Latinos as, 38
    rate, 25, 34–35, 37
    types of, 27
    against unauthorized immigrants, 31–32
    wealth of Americans draw, 26
Cruz, Emily Bonderer, 41–42
Cuban immigrants, 6

Davidson, Adam, 24
Day Without a Mexican, A (film), 11
deaths, 32
Deferred Action for Childhood Arrivals
    program (2012), 9, 64
Department of Justice (DOJ), 27–28
deportation
    of children of unauthorized immigrants, 43
    costs of, 43–44
    of criminals, 27
    Deferred Action for Childhood Arrivals
        and, 9, 64
    number of recent, 44
    public opinion about, 45
DREAM Act proposals, 9
DREAMers. See unauthorized immigrants
    under children
drug-traffickers, 25, 27–28, 38

economy
    areas of domination, 11
    competitiveness of American companies,
        20, 22
    contributions to tax base, 22–24
    cost of aging population to, 12, 23, 24
    dependence of, 12, 19–24, 64
    as draw for immigrants, 15
    harm to, 12, 13–18
    money invested in US, 23
    money sent out of US, 18, 22–23
    public opinion about effect of granting
        legal status, 45
    public opinion about impact on, 19–20
education system
    cost to taxpayers, 12, 16, 18
    dependence of, 21
    legal status and, 64
    Supreme Court decision, 17–18
El Paso, Texas, 33–34, **37**
El Shukrijumah, Adnan G., 31
employer sanctions, 51
E-Verify program, 64

federal government
    border enforcement approach of, is wrong,
        56–57
    Constitution leaves immigration policy
        to, 62
    cost of enforcement, 65
    cost of public services provided, 16
    has failed to enforce existing laws, 54, 64
    See also Border Patrol
Federation for American Immigration Reform
    (FAIR), 13, 14, 17
financial activities industry, **21**
Finger, Richard, 52
food costs to consumers, 15, 20
Forbes (magazine), 52

Gascon, George, 62
Georgia, 64
Goss, Stephen, 24
Griswold, Daniel, 33, 35–36

Hagel, Chuck, 56–57
Hinojosa, Raul, 22–23
hospitality industry, **21**
House Committee on Homeland Security, 31
Hsu, Spencer S., 36
human smuggling, 31

identity theft, 15
immigrants
    difficulties determining legal status of,
        63–64
    increase safety of cities, 25, 34, 36
    state laws drive wedge between law
        enforcement and, 62–63
    wait to become legal, 44, 46
Immigration and Customs Enforcement
    (ICE), 55
Immigration Policy Center (IPC), 34–35,
    46
Immigration Reform and Control Act
    (1986), 9, 51
imprisonment, 27
Individual Taxpayer Identification Numbers,
    23
industry, dependence of, **21**
information industry, **21**
investments by Latin Americans, 23
Islamic fundamentalism, 30–31

Jacobs, Gary, 47, 51–52

Khazanovich, Alex, 51
Kobach, Kris W., 54, 59

Larson, Erik, 20
law enforcement
    costs to federal government, 65
    should be by state and local authorities,
        53, 54–58
    should not be by state and local authorities,
        53, 61–66
    state and local authorities are already
        undertaking, 55, 58, 59–60
legal status, public opinion about effect of
    granting, **45**
Let Freedom Ring, 66
Levin, Jack, 34
Lipman, Francine J., 24
local government. *See* state and local
    government

manufacturing, **21**
Mara Salvatrucha gang (MS-13), 31
Martin, Philip, 15
McCain, John, 42–43
McGrath, Roger D., 26, 28
Mehlman, Ira, 60
Melisio, Maria Lola, 64

Mexico
    border area
        fence along, 49–50, 65–66
        is dangerous, 25
        is not dangerous, 25, 33–34, 35
        is used by "irregular migrants," 30
        US militarization is correct approach,
            65–66
        US militarization is wrong approach,
            56–57
    conditions in, 7–8
    drug-traffickers from, are primary US
        suppliers, 28
    economic support by unauthorized
        immigrants, 18
    number of visas available to citizens of, 46
    See also Border Patrol
militarization of border area, 56–57, 65–66
Minuteman Civil Defense Corps, 6, 55, 56
MS-13 gang, 31
Myers, Dowell, 46

National Drug Threat Assessment (2010),
    27–28
National Immigration Law Center, 41
New York Times (newspaper), 19, 22, 44, 46
Nogales, Arizona, 36
Norway, 22
Nowrasteh, Alex, 44
NumbersUSA, 55

Obama, Barack, 9, 44

Parker, Christina, 57
Peri, Giovanni, 22
Phoenix, Arizona, crime in, 28, 33, 63
Plyler v. Doe, 17–18
Policy Link, 63
Porter, Eduardo, 19, 22
Preston, Julia, 44, 46
public opinion
    impact of unauthorized immigrants on
        economy, 19–20
    legal status for unauthorized immigrants,
        45
    path to citizenship, 40, 42–43
public services
    aging population need for, 12, 23, 24, 40
    children of unauthorized immigrants will
        support, 23
    cost to taxpayers, 12, 16, 17, 50

jobs taken by unauthorized immigrants
force citizens on welfare, 14
public opinion about effect of granting
legal status on, 45
states requiring legal status for receipt of,
58
unauthorized immigrants help subsidize,
12, 23

Real Housewife of Ciudad Juárez, The (blog),
42
residency path for those brought as children,
9
Rumbaut, Ruben G., 34–35

Saenz, Thomas A., 63
safety and security
unauthorized immigrants do not threaten,
25, 33–39
unauthorized immigrants threaten, 25,
26–32, 29
SB 1070 (Arizona immigration law),
35–37
Secure Fence Act (2006), 49–50, 65–66
"show me your papers" provision (Arizona
law SB 1070), 35–37
Simcox, Chris, 56
smuggling
of drugs, 25, 27–28, 38
of humans, 31
Social Security
children of unauthorized immigrants will
support, 23
theft of numbers, 15
unauthorized immigrants do not use,
23–24
unauthorized immigrants help support,
12, 23–24
Spain, 49
Spurlock, Morgan, 6–7
state and local government
costs to, 12, 16, 18, 59–60
immigration laws
are already enforcing, 58, 59–60
should enforce, 53, 54–58
should not enforce, 53, 61–66
Stoddard, David, 66
Supreme Court decisions, 17–18, 35
Swift & Company, 15

tax base contributions, 22–24

terrorist threats
apprehensions by country of origin of
unauthorized immigrants, 29
border policies will not reduce, 38–39, 66
chances of, are increased, 25, 29–31
Texas
benefits granted to US-born children, 17
cost to, 60
crime rate in El Paso, 33–34, 37
Theodore, Nik, 63
30 Days (television program), 6–7
*Transnational Organized Crime in Central
America and the Caribbean* (UN Office on
Drugs and Crime), 30

unauthorized immigrants
changes in number of, 10–11, 65
major countries of origin of, 10
states with largest number of, 8, 10
total in US (January 2011), 8
total in US (2013), 10
United Nations Office on Drugs and Crime,
30
US-Mexico border
fence along, 49–50, 65–66
is dangerous area, 25
is not dangerous area, 25, 33–34, 35
is used by "irregular migrants," 30
US militarization is correct approach,
65–66
US militarization is wrong approach,
56–57
See also Border Patrol

wages, 13–14, 15, 22–23
Washington Post (newspaper), 36
workforce
aging population will need increase in, 23,
40, 46
areas are important in, 11, 20, 21
employer sanctions, 51
number in, 14
provide jobs for more skilled Americans, 22
public opinion about effect on, of granting
legal status, 45
take jobs Americans do not want, 12, 64
take jobs from Americans, 12, 13, 14
unskilled workers will be needed, 46

youth. *See* children
Yuma, Arizona, 49–50

About the Author

John Allen is a freelance writer in Oklahoma City, Oklahoma. He has worked in the educational publishing field for many years and has written books on a variety of topics.